SAFETY SYMBOLS

SAFETY SYMBOLS	HAZARD	EXAMPLES	PRECAUTION	REMEDY
DISPOSAL	Special disposal procedures need to be followed.	certain chemicals, living organisms	Do not dispose of these materials in the sink or trash can.	Dispose of wastes as directed by your teacher.
BIOLOGICAL	Organisms or other biological materials that might be harmful to humans	bacteria, fungi, blood, unpreserved tissues, plant materials	Avoid skin contact with these materials. Wear mask or gloves.	Notify your teacher if you suspect contact with material. Wash hands thoroughly.
EXTREME TEMPERATURE	Objects that can burn skin by being too cold or too hot	boiling liquids, hot plates, dry ice, liquid nitrogen	Use proper protection when handling.	Go to your teacher for first aid.
SHARP OBJECT	Use of tools or glassware that can easily puncture or slice skin	razor blades, pins, scalpels, pointed tools, dissecting probes, broken glass	Practice common-sense behavior and follow guidelines for use of the tool.	Go to your teacher for first aid.
FUME	Possible danger to respiratory tract from fumes	ammonia, acetone, nail polish remover, heated sulfur, moth balls	Make sure there is good ventilation. Never smell fumes directly. Wear a mask.	Leave foul area and notify your teacher immediately.
ELECTRICAL	Possible danger from electrical shock or burn	improper grounding, liquid spills, short circuits, exposed wires	Double-check setup with teacher. Check condition of wires and apparatus.	Do not attempt to fix electrical problems. Notify your teacher immediately.
IRRITANT	Substances that can irritate the skin or mucous membranes of the respiratory tract	pollen, moth balls, steel wool, fiberglass, potassium permanganate	Wear dust mask and gloves. Practice extra care when handling these materials.	Go to your teacher for first aid.
CHEMICAL	Chemicals that can react with and destroy tissue and other materials	bleaches such as hydrogen peroxide; acids such as sulfuric acid, hydrochloric acid; bases such as ammonia, sodium hydroxide	Wear goggles, gloves, and an apron.	Immediately flush the affected area with water and notify your teacher.
TOXIC	Substance may be poisonous if touched, inhaled, or swallowed	mercury, many metal compounds, iodine, poinsettia plant parts	Follow your teacher's instructions.	Always wash hands thoroughly after use. Go to your teacher for first aid.
OPEN FLAME	Open flame may ignite flammable chemicals, loose clothing, or hair	alcohol, kerosene, potassium permanganate, hair, clothing	Tie back hair. Avoid wearing loose clothing. Avoid open flames when using flammable chemicals. Be aware of locations of fire safety equipment.	Notify your teacher immediately. Use fire safety equipment if applicable.

Eye Safety Proper eye protection should be worn at all times by anyone performing or observing science activities.

Clothing Protection This symbol appears when substances could stain or burn clothing.

Animal Safety This symbol appears when safety of animals and students must be ensured.

Radioactivity This symbol appears when radioactive materials are used.

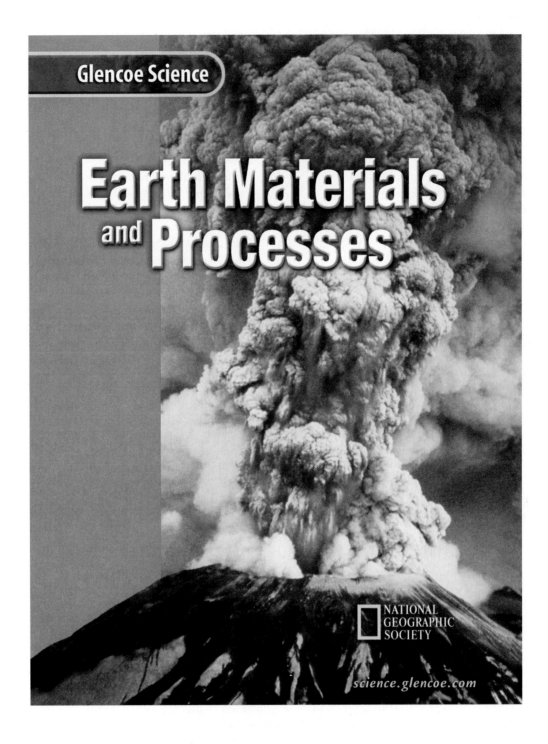

Glencoe Science

Earth Materials
and Processes

NATIONAL
GEOGRAPHIC
SOCIETY

science.glencoe.com

**Glencoe
McGraw-Hill**

New York, New York   Columbus, Ohio   Woodland Hills, California   Peoria, Illinois

Glencoe Science

Earth Materials and Processes

Student Edition
Teacher Wraparound Edition
Interactive Teacher Edition CD-ROM
Interactive Lesson Planner CD-ROM
Lesson Plans
Content Outline for Teaching
Dinah Zike's Teaching Science with Foldables
Directed Reading for Content Mastery
Foldables: Reading and Study Skills
Assessment
 Chapter Review
 Chapter Tests
 ExamView Pro Test Bank Software
 Assessment Transparencies
 Performance Assessment in the Science Classroom
 The Princeton Review Standardized Test Practice Booklet
Directed Reading for Content Mastery in Spanish
Spanish Resources
English/Spanish Guided Reading Audio Program
Reinforcement

Enrichment
Activity Worksheets
Section Focus Transparencies
Teaching Transparencies
Laboratory Activities
Science Inquiry Labs
Critical Thinking/Problem Solving
Reading and Writing Skill Activities
Mathematics Skill Activities
Cultural Diversity
Laboratory Management and Safety in the Science Classroom
MindJogger Videoquizzes and Teacher Guide
Interactive CD-ROM with Presentation Builder
Vocabulary PuzzleMaker Software
Cooperative Learning in the Science Classroom
Environmental Issues in the Science Classroom
Home and Community Involvement
Using the Internet in the Science Classroom

THE PRINCETON REVIEW

"Study Tip," "Test-Taking Tip," and the "Test Practice" features in this book were written by The Princeton Review, the nation's leader in test preparation. Through its association with McGraw-Hill, The Princeton Review offers the best way to help students excel on standardized assessments.

The Princeton Review is not affiliated with Princeton University or Educational Testing Service.

Glencoe/McGraw-Hill

A Division of The McGraw·Hill Companies

Cover Images: Eruption column above Mount St. Helens, Washington, during the 1980 eruption

Send all inquiries to:
Glencoe/McGraw-Hill
8787 Orion Place
Columbus, OH 43240

ISBN 0-07-825531-7
Printed in the United States of America.
4 5 6 7 8 9 10 027/043 06 05 04 03 02

Authors

National Geographic Society
Education Division
Washington, D.C.

Ralph M. Feather Jr., PhD
Science Department Chair
Derry Area School District
Derry, Pennsylvania

Dinah Zike
Educational Consultant
Dinah-Might Activities, Inc.
San Antonio, Texas

Content Consultants

William C. Keel, PhD
Department of Physics
and Astronomy
University of Alabama
Tuscaloosa, Alabama

Stephen M. Letro
National Weather Service
Meteorologist in charge
Jacksonville, Florida

Robert Nierste
Science Department Head
Hendrick Middle School
Plano, Texas

Series Safety Consultants

Malcolm Cheney, PhD
OSHA Chemical Safety
Officer
Hall High School
West Hartford, Connecticut

Aileen Duc, PhD
Science II Teacher
Hendrick Middle School
Plano, Texas

Sandra West, PhD
Associate Professor of Biology
Southwest Texas State
University
San Marcos, Texas

Series Math Consultants

Michael Hopper, D. Eng
Manager of Aircraft Certification
Raytheon Company
Greenville, Texas

Teri Willard, EdD
Department of Mathematics
Montana State University
Belgrade, Montana

Reading Consultant

Nancy Woodson, PhD
Professor of English
Otterbein College
Westerville, Ohio

Reviewers

Lois Burdette
Green Bank Elementary-Middle School
Green Bank, West Virginia

Marcia Chackan
Pine Crest School
Boca Raton, Florida

Mary Ferneau
Westview Middle School
Goose Creek, South Carolina

Connie Cook Fontenot
Bethune Academy
Houston, Texas

Annette Garcia
Kearney Middle School
Commerce City, Colorado

Nerma Coats Henderson
Pickerington Jr. High School
Pickerington, Ohio

Sharon Mitchell
William D. Slider Middle School
El Paso, Texas

Joanne Stickney
Monticello Middle School
Monticello, New York

Series Activity Testers

José Luis Alvarez, PhD
Math/Science Mentor Teacher
El Paso, Texas

Nerma Coats Henderson
Teacher
Pickerington Jr. High School
Pickerington, Ohio

Mary Helen Mariscal-Cholka
Science Teacher
William D. Slider Middle School
El Paso, Texas

José Alberto Marquez
TEKS for Leaders Trainer
El Paso, Texas

Science Kit and Boreal Laboratories
Tonawanda, New York

CONTENTS

CONTENTS

Interdisciplinary Connections/Activities

Activities/Science Connections

Problem-Solving Activities

Math Skills Activities

Skill Builder Activities

Feature Contents

Science
INTEGRATION

SCIENCE *Online*

THE
PRINCETON
REVIEW

Monitoring Voclanoes

Volcanic eruptions can cause incredible destruction, yet many people continue to live near active volcanoes. One approach to protect lives and property is to look for signs that a volcano is about to erupt. This was done on Mount St. Helens in the state of Washington prior to its eruption on May 18, 1980.

Mount St. Helens exploded after 123 years of inactivity. Over 600 km² of surrounding land was devastated. More than 300 m of the volcano's north face blew away, creating a huge crater and sending a cloud of hot steam and ash roaring down the flanks of its north slope.

On the island of Hawaii, the Mauna Loa and Kilauea volcanoes erupt more quietly than Mount St. Helens, but they still have the potential to cause great damage. In 1990, lava flows from Kilauea destroyed property in Kalapana Gardens. In 1984, an eruption of Mauna Loa sent lava to within 6.5 km of Hilo, the largest city on the island of Hawaii.

Figure 1
The May 18, 1980, eruption of Mount St. Helens blew tons of ash, rock, and steam into the air when it erupted.

Figure 2
The eruption of Mount St. Helens killed 57 people and caused hundreds of millions of dollars in damage. The force of the blast knocked down millions of trees.

Living Near a Volcano

Volcanoes are natural environmental hazards because of their potentially destructive power and their proximity to populated areas. Many people are reluctant or unwilling to move from their homes near active volcanoes even though there is no way to prevent volcanic eruptions. Such regions often enjoy rich soils of volcanic origin. Consequently, scientists have been working for many years to find the best ways to monitor various volcanoes around the world. They suggest that the data they gather will enable them to better forecast when a quiet volcano might erupt again, allowing people to evacuate a region before an eruption.

Figure 3
Kilauea has erupted continuously for more than 15 years. This lava flow encroached on property in Kalapana Gardens in 1990.

Science

Some advances in the study of volcanoes came about as scientists first attempted to solve the problem of how to forecast eruptions. Solving problems to help make people's lives safer and better is a benefit of science. When you solve a problem by finding a better way to do something, you are doing science.

Volcanology is part of Earth science, the scientific study of the solid part of Earth, the oceans, the atmosphere, and bodies in space. In this book, you will learn about the materials of which Earth is made. You also will learn about processes, such as volcanic eruptions, that shape and change Earth's surface.

Figure 4
Hilo, Hawaii, sits in the path of volcanic lava flows.

Science Today

For most of human history, volcanic eruptions have caught people off-guard. Eruptions have poured out lava, hot ash, and gas, often trapping people before they could escape. Today, although eruptions still cause great destruction, fewer people die because volcanologists—scientists who study volcanoes—can forecast many eruptions. For instance, workers knew that Mount St. Helens would explode thanks to advances in volcano monitoring techniques. They were able to warn people in the area and save many lives.

Looking for Signs

Monitoring is reading the signs of activity generated by a volcano before an eruption. For example, prior to a volcanic eruption, magma moves toward Earth's surface. This movement causes earthquakes, changes in a volcano's shape, and the release of certain gases. Volcanologists use specialized instruments to measure changes in the ground surface, the amounts and types of gases emitted, and seismic waves released by earthquakes.

One sign that a volcano might erupt is an increase in the number of earthquakes in the region. Magma and gases force their way up through cracks deep in a volcano, causing the earthquakes. For example, two months before the eruption of Mount St. Helens, about 10,000 quakes occurred in the mountain. Seismographs placed on or near volcanoes can record such earthquakes.

Volcanologists also know that changes in the shape of a volcano can mean an eruption might soon occur. As magma moves upward, parts of a volcano might rise or sink. Mount St. Helens formed a huge bulge in the weeks prior to its eruption.

Where Volcanologists Work

Some scientists who monitor volcanoes work at the United States Geological Survey (USGS) volcano observatories, such as:

1. **Alaska Volcano Observatory:** Monitors Alaska's volcanoes and sends out warnings about eruptions in eastern Russia.
2. **Hawaii Volcano Observatory:** Monitors the active volcanoes on the island of Hawaii.
3. **Cascades Volcano Observatory:** Monitors and assesses hazards from volcanoes of the Cascade Range.
4. **Long Valley Observatory:** Monitors activity from the large and potentially hazardous calderas system near Mammoth Lakes, California.

Using Technology

Besides seismographs, volcanologists use tiltmeters, electronic distance meters (EDMs), spectrometers, and strainmeters. A tiltmeter measures changes in the slope of the ground caused by moving magma. Like a carpenter's level, it consists of a bubble inside a fluid-filled container. If the slope changes, the bubble moves and the difference is measured electronically. An electronic distance meter uses a laser beam to measure the distance between two points on a volcano. If magma moves rocks or widens cracks, the targets will move and the EDM will record a change in distance.

Spectrometers measure gases released from magma. The rate at which volcanoes release carbon dioxide and sulfur dioxide, for example, might change before an eruption.

The strainmeter (or dilatometer) is being used in Hawaii to monitor Mauna Loa and Kilauea. It consists of a small canister filled with liquid silicon that is placed deep in a hole drilled into a volcano. Any movement in the volcano that changes the shape of the ground squeezes the strainmeter and the measurements are recorded on instruments at the surface.

Working on a Volcano

Although some volcanoes are monitored using radio-controlled instruments, volcanologists also must work in dangerous conditions on active volcanoes. They install instruments, take readings, or collect gas escaping from volcanic vents.

Volcanologist Cynthia Gardner enjoys her work in Washington, Oregon, and Alaska because she's helping to save lives. When she's not in the field, she collects data, writes reports, and sets up emergency procedures in communities near volcanoes.

Figure 5
This USGS solar-powered seismograph records small earthquakes on the flank of the Augustine volcano in Alaska.

Figure 6
Volcanologist Cynthia Gardner uses advanced equipment to monitor volcanoes.

You Do It

Airplanes and satellites are tools that help volcanologists forecast the eruption of volcanoes. Research in your local library or on the Glencoe Science Web site at **science.glencoe.com** to find out how volcanologists employ these tools. How would their work be more difficult without the aid of airplanes and satellites?

Minerals

These gorgeous diamonds appear so perfect that it seems impossible they formed naturally. Although it's true that these specimens were cut by humans to enhance their beauty, some minerals seem flawless the moment they're found. As you'll soon learn, a requirement for being a mineral is that it must be naturally occurring. In the case of a diamond, this increases its rarity and its price. In this chapter, you'll learn about minerals and gems and how to identify them. You'll also learn how much society depends upon minerals.

What do you think?

Science Journal Look at the picture below with a classmate. Discuss what this might be. Here's a hint: *This "egg" was laid by a volcano.* Write your answer or best guess in your Science Journal.

EXPLORE ACTIVITY

What's the difference between a rock and a mineral? Just as a building is made of many different materials such as concrete, wood, plaster, and steel, rocks are made of many different things. Most rocks are made from crystals of several different minerals. Some rocks are made from many different crystals of the same mineral. When examining a rock, you'll notice that it is made from many different components. However, a mineral looks the same throughout. Can you tell a rock from a mineral?

Distinguish rocks from minerals

1. Use a magnifying glass to examine a quartz crystal, salt grains, and samples of sandstone, granite, calcite, mica, and schist (SHIHST).

2. Determine which samples are made of one type of material and should be classified as minerals.

3. Determine which samples are made of more than one type of material and should be classified as rocks.

Observe

In your Science Journal, compile a list of descriptions for the minerals you examined and a second list of descriptions for the rocks. Compare and contrast your observations of minerals and rocks.

Before You Read

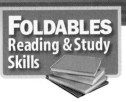

FOLDABLES
Reading & Study Skills

Making a Question Study Fold Asking yourself questions helps you stay focused and better understand minerals when you are reading the chapter.

1. Place a sheet of notebook paper in front of you so the short side is at the top and the holes are on the right side. Fold the paper in half from the left side to the right side.

2. Through the top thickness of paper, cut along every third line from the outside edge to the center fold, forming tabs as shown.

3. Before you read the chapter, write questions you have about minerals on the front of the tabs. As you read the chapter, add more questions and write answers under the tabs.

Minerals

What You'll Learn
- **Describe** characteristics that all minerals share.
- **Explain** how minerals form.

Vocabulary
mineral
crystal
magma
silicate

Why It's Important
You use minerals every day.

What is a mineral?

How important are minerals to you? Very important? You actually own or encounter many things made from minerals every day. Ceramic, metallic, and even some paper items are examples of products that are derived from or include minerals. **Figure 1** shows just a few of these things. Metal bicycle racks, bricks, and the glass in windows would not exist if it weren't for minerals. A **mineral** is a naturally occurring, inorganic solid with a definite chemical composition and an orderly arrangement of atoms. About 4,000 different minerals are found on Earth, but they all share these four characteristics.

Mineral Characteristics First, all minerals are formed by natural processes. These are processes that occur on or inside Earth with no input from humans. For example, salt formed by the natural evaporation of seawater is the mineral halite, but salt formed by evaporation of saltwater solutions in laboratories is not a mineral. Second, minerals are inorganic. This means that they aren't made by life processes. Third, every mineral is an element or compound with a definite chemical composition. For example, halite's composition, NaCl, gives it a distinctive taste that adds flavor to many foods. Fourth, minerals are crystalline solids. All solids have a definite volume and shape. Gases and liquids like air and water have no definite shape, and they aren't crystalline. Only a solid can be a mineral, but not all solids are minerals.

Atom Patterns The word *crystalline* means that atoms are arranged in a pattern that is repeated over and over again. For example, graphite's atoms are arranged in layers. Opal, on the other hand, is not a mineral in the strictest sense because its atoms are not all arranged in a definite, repeating pattern, even though it is a naturally occurring, inorganic solid.

Figure 1
You probably use minerals or materials made from minerals every day without thinking about it. *How many objects in this picture might be made from minerals?*

Figure 2
More than 200 years ago, the smooth, flat surfaces on crystals led scientists to infer that minerals had an orderly, internal structure inside.

B

A

A The well-formed crystal shapes exhibited by these clear quartz crystals suggest an orderly structure. B Even though this rose quartz looks uneven on the outside, its atoms have an orderly arrangement on the inside.

The Structure of Minerals

Do you have a favorite mineral sample or gemstone? If so, perhaps it contains well-formed crystals. A **crystal** is a solid in which the atoms are arranged in orderly, repeating patterns. You can see evidence for this orderly arrangement of atoms when you observe the smooth, flat outside surfaces of crystals. A crystal system is a group of crystals that have similar atomic arrangements and therefore similar external crystal shapes.

☑ **Reading Check**   *What is a crystal?*

Crystals Not all mineral crystals have smooth surfaces and regular shapes like the clear quartz crystals in **Figure 2A.** The rose quartz in **Figure 2B** has atoms arranged in repeating patterns, but you can't see the crystal shape on the outside of the mineral. This is because the rose quartz crystals developed in a tight space, while the clear quartz crystals developed freely in an open space. The six-sided, or hexagonal crystal shape of the quartz crystals in **Figure 2A,** and other forms of quartz can be seen in some samples of the mineral. **Figure 3** illustrates the six major crystal systems, which classify minerals according to their crystal structures. The hexagonal system to which quartz belongs is one example of a crystal system.

Crystals form by many processes. Next, you'll learn about two of these processes—crystals that form from magma and crystals that form from solutions of salts.

Inferring Salt's Crystal System

Procedure
1. Use a **magnifying glass** and a **dissecting probe** to observe grains of common **table salt.** Sketch the shape of a salt grain. **WARNING:** *Do not taste or eat mineral samples. Keep hands away from your face.*
2. Compare the shapes of the salt crystals with the shapes of crystals shown in **Figure 3.**

Analysis
1. Which characteristics do all the grains have in common?
2. Research another mineral with the same crystal system as salt. What is this crystal system called?

Figure 3

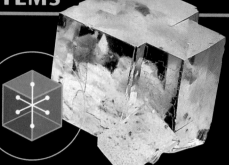

A crystal's shape depends on how its atoms are arranged. Crystal shapes can be organized into groups known as crystal systems—shown here in 3-D with geometric models (in blue). Knowing a mineral's crystal system helps researchers understand its atomic structure and physical properties.

▲ CUBIC Fluorite is an example of a mineral that forms cubic crystals. Minerals in the cubic crystal system are equal in size along all three principal dimensions.

▲ HEXAGONAL (hek SA guh nul) In hexagonal crystals, horizontal distances between opposite crystal surfaces are equal. These crystal surfaces intersect to form 60° or 120° angles. The vertical length is longer or shorter than the horizontal lengths.

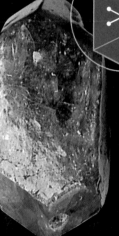

◀ TETRAGONAL (te TRA guh nul) Zircon crystals are tetragonal. Tetragonal crystals are much like cubic crystals, except that one of the principal dimensions is longer or shorter than the other two dimensions.

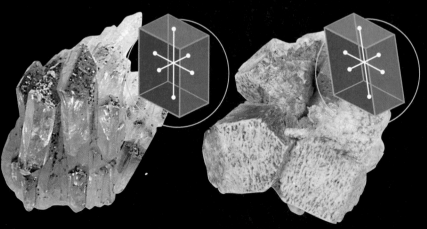

▲ ORTHORHOMBIC (awr thuh RAHM bihk) Minerals with orthorhombic structure, such as barite, have dimensions that are unequal in length, resulting in crystals with a brick-like shape.

▲ MONOCLINIC (mah nuh KLIH nihk) Minerals in the monoclinic system, such as orthoclase, also exhibit unequal dimensions in their crystal structure. Only one right angle forms where crystal surfaces meet. The other angles are oblique, which means they don't form 90° angles where they intersect.

▲ TRICLINIC (tri KLIH nihk) The triclinic crystal system includes minerals exhibiting the least symmetry. Triclinic crystals, such as rhodonite (ROH dun ite), are unequal in all dimensions, and all angles where crystal surfaces meet are oblique.

Figure 4
Minerals form by many natural processes.

A This rock formed as magma cooled slowly, allowing large mineral grains to form.

Labradorite

B Some minerals form when salt water evaporates, such as these white crystals of halite in Death Valley, California.

Crystals from Magma
Natural processes form minerals in many ways. For example, hot melted rock, called **magma,** cools when it reaches Earth's surface, or even if it's trapped below the surface. As magma cools, its atoms lose heat energy, move closer together, and begin to combine into compounds. Molecules of the different compounds then arrange themselves into orderly, repeating patterns. The type and amount of elements present in a magma partly determine which minerals will form. Also, the size of the crystals that form depends partly on how rapidly the magma cools.

When magma cools slowly, the crystals that form are generally large enough to see with the unaided eye, as shown in **Figure 4A.** This is because the atoms have enough time to move together and form into larger crystals. When magma cools rapidly, the crystals that form will be small. In such cases, you can't easily see individual mineral crystals.

Crystals from Solution
Crystals also can form from minerals dissolved in water. When water evaporates, as in a dry climate, ions that are left behind can come together to form crystals like those in **Figure 4B.** Or, if too much of a substance is dissolved in water, ions can come together and crystals of that substance can begin to form in the solution. Minerals can come out of a solution in this way without the need for evaporation.

Physics
INTEGRATION
Evaporites commonly form in dry climates. Research the change that takes place when a saline lake or shallow sea evaporates and halite or gypsum forms.

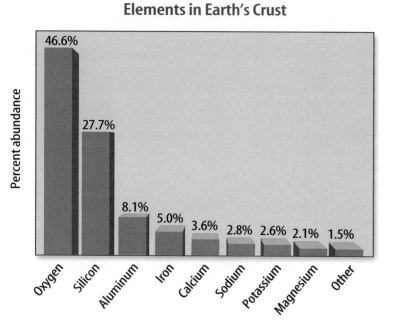

Elements in Earth's Crust

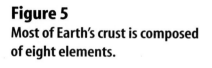

Figure 5
Most of Earth's crust is composed of eight elements.

Mineral Compositions and Groups

Ninety elements occur naturally in Earth's crust. Approximately 98 percent (by weight) of the crust is made of only eight of these elements, as in **Figure 5.** Of the thousands of known minerals, only a few dozen are common, and these are mostly composed of the eight common elements in Earth's crust.

Most of the common rock-forming minerals belong to a group called the silicates. **Silicates** (SIH luh kayts) are minerals that contain silicon (Si) and oxygen (O) and usually one or more other elements. Silicon and oxygen are the two most abundant elements in Earth's crust. These two elements alone combine to form the basic building blocks of most of the minerals in Earth's crust and mantle. Feldspar and quartz, which are silicates, and calcite, which is a carbonate, are examples of common, rock-forming minerals. Other mineral groups also are defined according to their compositions.

Section 1 Assessment

1. What four characteristics must a substance have to be a mineral?

2. Describe two processes involving solutions that form minerals.

3. Diamonds can be made in the laboratory by subjecting starting materials containing carbon to high pressure. Are these diamonds minerals? Explain.

4. How are crystals of minerals classified?

5. **Think Critically** The mineral dolomite, a rock-forming mineral, contains oxygen, carbon, magnesium, and calcium. Is dolomite a silicate? In your Science Journal, compare and contrast rock-forming minerals and silicates.

Skill Builder Activities

6. **Classifying** Discover some items in your home or classroom that are made from minerals. How many items are metals? How many are non-metals? **For more help, refer to the** Science Skill Handbook.

7. **Using an Electronic Spreadsheet** Use spreadsheet software to make a graph of your own design that shows the relative percentages of the eight most common elements in Earth's crust. Using these eight elements, about what percentage of the crust is made up of iron and aluminum? Use the spreadsheet to perform this calculation. **For more help, refer to the** Technology Skill Handbook.

Activity

Crystal Formation

So far in this chapter, you've learned about minerals and how they form. In this activity, you'll have a chance to learn how crystals form from solutions.

What You'll Investigate
How do crystals form from solution?

Materials
250-mL beakers (2)	cotton string
cardboard	hot plate
large paper clip	hand lens
table salt	thermal mitt
flat wooden stick	shallow pan
granulated sugar	spoon

Goals
- **Compare and contrast** the crystals that form from salt and sugar solutions.
- **Observe** crystals and infer how they formed.

Safety Precautions

WARNING: *Never taste or eat any lab materials.*

Procedure

1. Gently mix separate solutions of salt in water and sugar in water in the two beakers. Keep stirring the solutions as you add salt or sugar to the water. Stop mixing when no more salt or sugar will dissolve in the solutions. Label each beaker as a salt or sugar solution.

2. Place the sugar beaker on a hot plate. Use the hot plate to heat the sugar solution gently. **WARNING:** *The liquid is hot. Do not touch the beaker without protecting your hands.*

3. Tie one end of the thread to the middle of the wooden stick. Tie a large paper clip to the free end of the string for weight. Place the stick across the opening of the sugar beaker so the thread dangles in the sugar solution.

4. Remove the beaker from the hot plate and cover it with cardboard. Place it in a location where it won't be disturbed.

5. Pour a thin layer of the salt solution into the shallow pan.

6. Leave the beaker and the shallow pan undisturbed for at least one week.

7. After one week, examine each solution with a hand lens to see whether crystals have formed.

Conclude and Apply

1. **Compare and contrast** the crystals that formed from the salt and the sugar solutions. How do they compare with samples of table salt and sugar?

2. **Describe** what happened to the saltwater solution in the shallow pan.

3. Did this same process occur in the sugar solution? Explain.

*C*ommunicating Your Data

Make a poster that describes your methods of growing salt and sugar crystals. Present your results to your class.

Mineral Identification

As You Read

***What* You'll Learn**

- **Describe** physical properties used to identify minerals.
- **Identify** minerals using physical properties such as hardness and streak.

Vocabulary

hardness	streak
luster	cleavage
specific gravity	fracture

***Why* It's Important**

Identifying minerals helps you recognize valuable mineral resources.

Physical Properties

Why can you recognize a classmate when you see him or her in a crowd away from school? A person's height or the shape of his or her face helps you tell that person from the rest of your class. Height and facial shape are two properties unique to individuals. Individual minerals also have unique properties that distinguish them.

Mineral Appearance Just like height and facial characteristics help you recognize someone, mineral properties can help you recognize and distinguish minerals. Color and appearance are two obvious clues that can be used to identify minerals.

However, these clues alone aren't enough to recognize most minerals. The minerals pyrite and gold are gold in color and can appear similar, as shown in **Figure 6.** As a matter of fact, pyrite often is called fool's gold. Gold is worth a lot of money, whereas pyrite has little value. You need to look at other properties of minerals to tell them apart. Some other properties to study include how hard a mineral is, how it breaks, and its color when crushed into a powder. Every property you observe in a mineral is a clue to its identity.

Figure 6
The general appearance of a mineral often is not enough to identify it.

Pyrite

Gold

A Using only color, observers can be fooled when trying to distinguish between pyrite and gold.

B The mineral azurite is identified readily by its striking blue color.

Hardness A measure of how easily a mineral can be scratched is its **hardness.** The mineral talc is so soft you can scratch it loose with your fingernail. Talcum powder is made from this soft mineral. Diamonds, on the other hand, are the hardest mineral. Some diamonds are used as cutting tools, as shown in **Figure 7.** A diamond can be scratched only by another diamond. Diamonds can be broken, however.

Reading Check *Why is hardness sometimes referred to as scratchability?*

Sometimes the concept of hardness is confused with whether or not a mineral will break. It is important to understand that even though a diamond is extremely hard, it can shatter if given a hard enough blow in the right direction along the crystal.

Mohs Scale In 1824, the Austrian scientist Friedrich Mohs developed a list of common minerals to compare their hardnesses. This list is called Mohs scale of hardness, as seen in **Table 1.** The scale lists the hardness of ten minerals. Talc, the softest mineral, has a hardness value of one, and diamond, the hardest mineral, has a value of ten.

Here's how the scale works. Imagine that you have a clear or whitish-colored mineral that you know is either fluorite or quartz. You try to scratch it with your fingernail and then with an iron nail. You can't scratch it with your fingernail but you can scratch it with the iron nail. Because the hardness of your fingernail is 2.5 and that of the iron nail is 4.5, you can determine the unknown mineral's hardness to be somewhere around 3 or 4. Because it is known that quartz has a hardness of 7 and fluorite has a hardness of 4, the mystery mineral must be fluorite.

Some minerals have a hardness range rather than a single hardness value. This is because atoms are arranged differently in different directions in their crystal structures.

Figure 7
Some saw blades have diamonds embedded in them to help slice through materials, such as this limestone. Blades are kept cool by running water over them.

Table 1 Mineral Hardness			
Mohs Scale	**Hardness**	**Hardness of Common Objects**	
Talc (softest)	1		
Gypsum	2	fingernail	(2.5)
Calcite	3	piece of copper	(2.5 to 3.0)
Fluorite	4	iron nail	(4.5)
Apatite	5	glass	(5.5)
Feldspar (orthoclase)	6	steel file	(6.5)
Quartz	7	streak plate	(7.0)
Topaz	8		
Corundum	9		
Diamond (hardest)	10		

Graphite

Fluorite

Luster Luster is the way a mineral reflects light. Luster can be metallic or nonmetallic. Minerals with a metallic luster, like the graphite shown in **Figure 8,** shine like metal. Metallic luster can be compared to the shine of a metal belt buckle, the shiny chrome trim on some cars, or the shine of metallic cooking utensils. When a mineral does not shine like metal, its luster is nonmetallic. Examples of terms for nonmetallic luster include *dull, pearly, silky,* and *glassy.* Common examples of minerals with glassy luster are quartz, calcite, halite, and fluorite.

Figure 8
Luster is an important physical property that is used to distinguish minerals. Graphite has a metallic luster. Fluorite has a nonmetallic, glassy luster.

Specific Gravity Minerals also can be distinguished by comparing their heft. The **specific gravity** of a mineral is the ratio of its weight compared with the weight of an equal volume of water. Like hardness, specific gravity is expressed as a number. If you were to research the specific gravities of gold and pyrite, you'd find that gold's specific gravity is about 17, and pyrite's is 5. This means that gold is about 17 times heavier than water and pyrite is 5 times heavier than water. You could experience this by comparing equal-sized samples of gold and pyrite in your hands—the pyrite would feel much lighter.

Problem-Solving Activity

How can you identify minerals?

Properties of Minerals		
Mineral	**Hardness**	**Streak**
Copper	2.5–3	copper-red
Galena	2.5	dark gray
Gold	2.5–3	yellow
Hematite	5.5–6.5	red to brown
Magnetite	6–6.5	black
Silver	2.5–3	silver-white

You have learned that minerals are identified by their physical properties, such as streak, hardness, cleavage, and color. Use your knowledge of mineral properties and your ability to read a table to solve the following problems.

Identifying the Problem
The table includes hardnesses and streak colors for several minerals. How can you use these data to distinguish minerals?

Solving the Problem
1. What test would you perform to distinguish hematite from copper? How would you carry out this test?
2. How could you distinguish copper from galena? What tool would you use?
3. What would you do if two minerals had the same hardness and the same streak color?

Streak When a mineral is rubbed across a piece of unglazed porcelain tile, as in **Figure 9,** a streak of powdered mineral is left behind. **Streak** is the color of a mineral when it is in a powdered form. The streak test works only for minerals that are softer than the streak plate. Gold and pyrite can be distinguished by a streak test. Gold has a yellow streak and pyrite has a greenish-black or brownish-black streak.

Some soft minerals will leave a streak even on paper. The last time you used a pencil to write on paper, you left a streak of the mineral graphite. One reason that graphite is used in pencil lead is because it is soft enough to leave a streak on paper.

Reading Check *Why do gold and pyrite leave a streak, but quartz does not?*

Figure 9
Streak is more useful for mineral identification than is mineral color. Hematite, for example, can be dark red, gray, or silver in color. However, its streak is always dark reddish-brown.

Cleavage and Fracture The way a mineral breaks is another clue to its identity. Minerals that break along smooth, flat surfaces have **cleavage** (KLEE vihj). Cleavage, like hardness, is determined by the arrangement of the mineral's atoms. Mica is a mineral that has one perfect cleavage. You can see in **Figure 10** how it breaks along smooth, flat planes. If you were to take a layer cake and separate its layers, you would show that the cake has cleavage. Not all minerals have cleavage. Minerals that break with uneven, rough, or jagged surfaces have **fracture.** Quartz is a mineral with fracture. If you were to grab a chunk out of the side of that cake, it would be like breaking a mineral that has fracture.

Mica

Halite

Figure 10
Weak bonds within the structures of mica and halite allow them to be broken along smooth, flat cleavage planes. *If you broke quartz, would it look the same?*

Mini LAB

Observing Mineral Properties

Procedure

1. Obtain samples of some of the following clear minerals: **gypsum, muscovite mica, halite,** and **calcite.**
2. Place each sample over the print on this page and observe the letters.

Analysis

1. Which mineral can be identified by observing the print's double image?
2. What other special property is used to identify this mineral?

Figure 11
Some minerals are natural magnets, such as this lodestone, which is a variety of magnetite.

Other Properties Some minerals have unique properties. Magnetite, as you can guess by its name, is attracted to magnets. Lodestone, a form of magnetite, will pick up iron filings like a magnet, as shown in **Figure 11.** Light forms two separate rays when it passes through some calcite specimens, causing you to see a double image. Calcite also can be identified because it fizzes when hydrochloric acid is put on it.

Now you know that you sometimes need more information than color and appearance to identify a mineral. You also might need to test its streak, hardness, luster, and cleavage or fracture. Although the overall appearance of a mineral can be different from sample to sample, its physical properties remain the same.

Section 2 Assessment

1. What is the difference between a mineral that has cleavage and one that has fracture?
2. How can an unglazed porcelain tile be used to identify a mineral?
3. Why is streak often more useful for mineral identification than color?
4. What hardness does a mineral have if it does not scratch glass but it scratches an iron nail?
5. **Think Critically** What does the presence of cleavage planes within a mineral tell you about the chemical bonds that hold the mineral together?

Skill Builder Activities

6. **Drawing Conclusions** A large piece of the mineral halite (salt) is broken repeatedly into several perfect cubes. How can this be explained? **For more help, refer to the** Science Skill Handbook.

7. **Making and Using Tables** In your Science Journal, make a list of three minerals. Next to each mineral, write another mineral that is similar in appearance. In a third column, list one physical property that can be used to distinguish the pair of similar-looking minerals. **For more help, refer to the** Science Skill Handbook.

③ Uses of Minerals

Gems

Walking past the window of a jewelry store, you notice a large selection of beautiful jewelry—a watch sparkling with diamonds, a necklace holding a brilliant red ruby, and a gold ring. For thousands of years, people have worn and prized minerals in their jewelry. What makes some minerals special? What unusual properties do they have that make them so valuable?

Properties of Gems Gems or gemstones, like the ones shown in **Figure 12,** are highly prized minerals because they are rare and beautiful. Most gems are special varieties of a particular mineral. They are clearer, brighter, or more colorful than common samples of that mineral. The difference between a gem and the common form of the same mineral can be slight. Amethyst is a gem form of quartz that contains just traces of iron in its structure. This small amount of iron gives amethyst a desirable purple color. Sometimes a gem has a crystal structure that allows it to be cut and polished to a higher quality than that of a non-gem mineral. **Table 2** lists popular gems and some locations where they have been collected.

As You Read

***What* You'll Learn**
■ **Describe** characteristics of gems that make them more valuable than other minerals.
■ **Identify** useful elements that are contained in minerals.

Vocabulary
gem

***Why* It's Important**
Minerals are necessary materials for decorative items and many manufactured products.

Figure 12
It is easy to see why gems are prized for their beauty and rarity. Shown here is The Imperial State Crown, made for Queen Victoria of England in 1838. It contains thousands of jewels, including diamonds, rubies, sapphires, and emeralds.

Table 2 Minerals and Their Gems

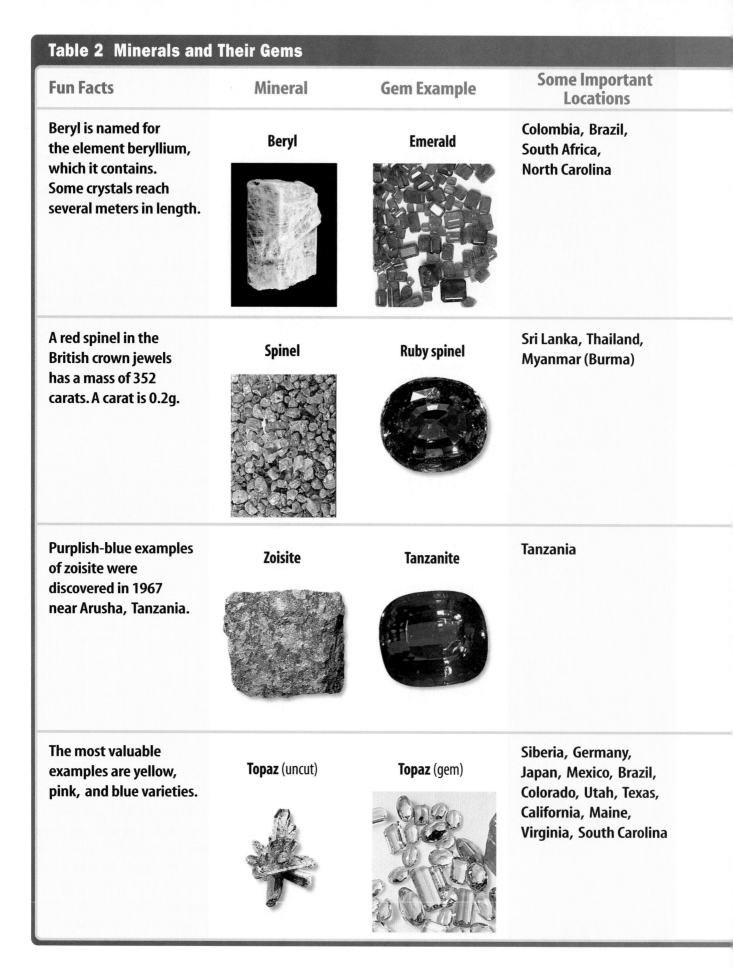

Fun Facts	Mineral	Gem Example	Some Important Locations
Beryl is named for the element beryllium, which it contains. Some crystals reach several meters in length.	**Beryl**	**Emerald**	Colombia, Brazil, South Africa, North Carolina
A red spinel in the British crown jewels has a mass of 352 carats. A carat is 0.2g.	**Spinel**	**Ruby spinel**	Sri Lanka, Thailand, Myanmar (Burma)
Purplish-blue examples of zoisite were discovered in 1967 near Arusha, Tanzania.	**Zoisite**	**Tanzanite**	Tanzania
The most valuable examples are yellow, pink, and blue varieties.	**Topaz** (uncut)	**Topaz** (gem)	Siberia, Germany, Japan, Mexico, Brazil, Colorado, Utah, Texas, California, Maine, Virginia, South Carolina

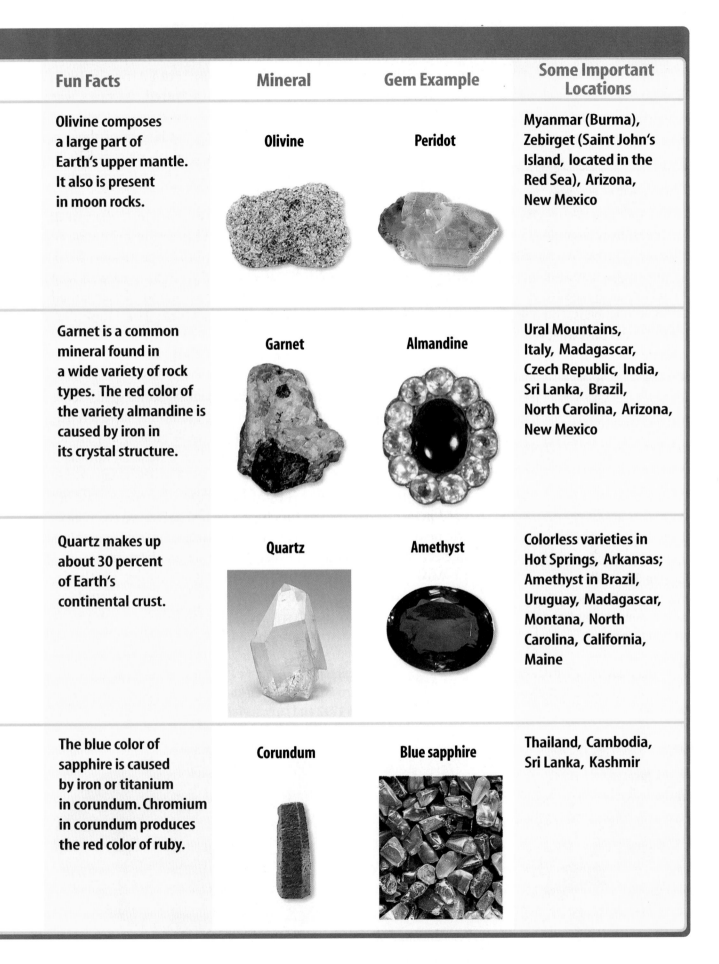

Fun Facts	Mineral	Gem Example	Some Important Locations
Olivine composes a large part of Earth's upper mantle. It also is present in moon rocks.	**Olivine**	**Peridot**	Myanmar (Burma), Zebirget (Saint John's Island, located in the Red Sea), Arizona, New Mexico
Garnet is a common mineral found in a wide variety of rock types. The red color of the variety almandine is caused by iron in its crystal structure.	**Garnet**	**Almandine**	Ural Mountains, Italy, Madagascar, Czech Republic, India, Sri Lanka, Brazil, North Carolina, Arizona, New Mexico
Quartz makes up about 30 percent of Earth's continental crust.	**Quartz**	**Amethyst**	Colorless varieties in Hot Springs, Arkansas; Amethyst in Brazil, Uruguay, Madagascar, Montana, North Carolina, California, Maine
The blue color of sapphire is caused by iron or titanium in corundum. Chromium in corundum produces the red color of ruby.	**Corundum**	**Blue sapphire**	Thailand, Cambodia, Sri Lanka, Kashmir

Important Gems All gems are prized, but some are truly spectacular and have played an important role in history. For example, the Cullinan diamond, found in South Africa in 1905, was the largest uncut diamond ever discovered. Its mass was 3,106.75 carats (about 621 g). The Cullinan diamond was cut into 9 main stones and 96 smaller ones. The largest of these is called the Cullinan 1 or Great Star of Africa. It's mass is 530.20 carats (about 106 g), and it is now part of the British monarchy's crown jewels, shown in **Figure 13A.**

Another well-known diamond is the blue Hope diamond, shown in **Figure 13B.** This is perhaps the most notorious of all diamonds. It was purchased by Henry Philip Hope around 1830, after whom it is named. Because his entire family as well as a later owner suffered misfortune, the Hope diamond has gained a reputation for bringing its owner bad luck. The Hope diamond's mass is 45.52 carats (about 9 g). Currently it is displayed in the Smithsonian Institution in Washington.

Useful Gems In addition to their beauty, some gems serve useful purposes. You learned earlier that diamonds have a hardness of 10 on Mohs scale. They can scratch almost any material—a property that makes them useful as industrial abrasives and cutting tools. Other useful gems include rubies, which are used to produce specific types of laser light. Quartz crystals are used in electronics and as timepieces. When subjected to an electric field, quartz vibrates steadily, which helps control frequencies in electronic devices and allows for accurate timekeeping.

Most industrial diamonds and other gems are synthetic, which means that humans make them. However, the study of natural gems led to their synthesis, allowing the synthetic varieties to be used by humans readily.

Figure 13
These gems are among the most famous examples of precious stones.

A The Great Star of Africa is part of a sceptre in the collection of British crown jewels.

B Beginning in 1668, the Hope diamond was part of the French crown jewels. Then known as the French Blue, it was stolen in 1792 and later surfaced in London, England in 1812.

Bauxite

Useful Elements in Minerals

Gemstones are perhaps the best-known use of minerals, but they are not the most important. Look around your home. How many things made from minerals can you name? Can you find anything made from iron?

Ores Iron, used in everything from frying pans to ships, is obtained from its ore, hematite. A mineral or rock is an ore if it contains a useful substance that can be mined at a profit. Magnetite is another mineral that contains iron.

Reading Check *When is a mineral also an ore?*

Extracting Elements Aluminum sometimes is refined, or purified, from the ore bauxite, shown in **Figure 14.** In the process of refining aluminum, aluminum oxide powder is separated from unwanted materials that are present in the original bauxite. After this, the aluminum oxide powder is converted to molten aluminum by a process called smelting.

During smelting, a substance is melted to separate it from any unwanted materials that may remain. Aluminum can be made into useful products like bicycles, soft-drink cans, foil, and lightweight parts for airplanes and cars. The plane flown by the Wright brothers during the first flight at Kitty Hawk had an engine made partly of aluminum.

Chemistry
INTEGRATION

In 1886, Charles Martin Hall discovered an important process that led to modern aluminum refinement. Research the Hall process. Describe what you learn in your Science Journal.

Figure 15
The mineral sphalerite (black) is an important source of zinc. Iron often is coated with zinc to prevent rust in a process called galvanization.

Vein Minerals Under certain conditions, metallic elements can dissolve in fluids. These fluids then travel through weaknesses in rocks and form mineral deposits. Weaknesses in rocks include natural fractures or cracks, faults, and surfaces between layered rock formations. Mineral deposits left behind that fill in the open spaces created by the weaknesses are called vein mineral deposits.

✔ Reading Check *How do fluids move through rocks?*

Sometimes, vein mineral deposits fill in the empty spaces after rocks collapse. An example of a mineral that can form in this way is shown in **Figure 15.** This is the shiny mineral sphalerite, a source of the element zinc, which is used in batteries. Sphalerite sometimes fills spaces in collapsed limestone.

Minerals Containing Titanium You might own golf clubs with titanium shafts or a racing bicycle containing titanium. Perhaps you know someone who has a titanium hip or knee replacement. Titanium is a durable, nontoxic, lightweight, metallic element derived from minerals that contain this metal in their atomic structures. Two minerals that are sources of the element titanium are ilmenite (IHL muh nite) and rutile (rew TEEL), shown in **Figure 16.** Ilmenite and rutile are common in rocks that form when magma cools and solidifies. They also occur as vein mineral deposits and in beach sands.

Figure 16
Rutile and ilmenite are common ore minerals of the element titanium.

Rutile

Ilmenite

Uses for Titanium Titanium is used in automobile body parts, such as connecting rods, valves, and suspension springs. It is used in the manufacture of aircraft and eyeglass frames. Low density and durability make titanium valuable in the production of these goods and of sports equipment such as tennis rackets and bicycles. Wheelchairs used by people who want to race or play basketball often are made from titanium, as shown in **Figure 17.** Titanium is one of many examples of useful materials that come from minerals and that enrich humans' lives.

Figure 17
Wheelchairs used for racing and playing basketball often have parts made from titanium.

Section 3 Assessment

1. Why is the Cullinan diamond considered to be important?
2. What do rubies and sapphires have in common?
3. Describe how vein minerals form.
4. Why is bauxite considered to be a useful rock?
5. **Think Critically** Titanium is a nontoxic metal. Why is this an important characteristic in the manufacture of artificial body parts?

Skill Builder Activities

6. **Comparing and Contrasting** Compare and contrast gem-quality amethyst with regular quartz. **For more help,** refer to the Science Skill Handbook.
7. **Using Percentages** On average, Earth's continental crust contains 5 percent iron and 0.007 percent zinc. How many times more iron than zinc is present in the average continental crust? **For more help,** refer to the Math Skill Handbook.

Mineral Identification

S ome mineral properties can be more useful than others when testing a particular mineral. Although certain minerals can be identified observing only one property, others require testing several properties to identify them.

Recognize the Problem

How can you identify unknown minerals?

Form a Hypothesis

Make a hypothesis about which properties you think will be most useful in identifying particular minerals.

Goals

- **Determine** which properties of a mineral you will use for identification purposes.
- **Design** an experiment that tests these properties to help identify minerals.

Possible Materials

mineral samples	streak plate
hand lens	5% hydrochloric
pan balance	acid (HCl) with
graduated cylinder	dropper
water	*vinegar
piece of copper	Mohs scale
*copper penny	of hardness
glass plate	Minerals Appendix
small iron nail	safety goggles
steel file	*Alternate materials

Safety Precautions

Use care when handling acids. Be careful when handling materials with sharp edges. **WARNING:** *HCl can cause burns. If spillage occurs, notify your teacher and rinse with a generous flow of cool water until you are told to stop. Do not taste, eat, or drink any lab materials. Never hold a glass plate in your hands while testing hardness because the plate could break.*

Test Your Hypothesis

Plan

1. As a group, agree upon and write your hypothesis statement.

2. As a group, list the steps you will need to take to test your hypothesis. Be specific, describing exactly what you will do at each step.

3. **List** the materials you will need to complete your experiment.

4. **List** the various properties of minerals that you will test.

5. **Predict** any special properties you expect to observe or test.

6. Reread your experiment to make sure that steps are in a logical order.

7. Should you test the properties of the minerals more than once?

8. How will you summarize your data?

9. **Identify** all constants, variables, and controls of the experiment.

Do

1. Make sure your teacher approves your plan before you start.

2. Carry out the experiment as planned.

3. While doing the experiment, write any observations that you or other members of your group make. Summarize your data in your Science Journal.

Analyze Your Data

1. Which properties were most useful in identifying your samples? Which properties were least useful?

2. **Compare** the properties that worked best for you with those that worked best for other students.

Draw Conclusions

1. For five minerals, discuss reasons why certain properties are useful and others are not.

2. **Determine** two properties that distinguish clear, transparent quartz from clear, transparent calcite. Explain your choice of properties.

*C*ommunicating
Your Data

For three minerals, **list** physical properties that were important for their identification. **For more help, refer to the** Science Skill Handbook.

Dr. Dorothy

Like X rays, electrons are diffracted by crystalline substances, revealing information about their internal structures and symmetry. This electron diffraction pattern of titanium was obtained with an electron beam focused along a specific direction in the crystal.

What contributions did Dorothy Crowfoot Hodgkin make to science?

Dr. Hodgkin used a method called X-ray crystallography (kris tuh LAH gruh fee) to figure out the structures of crystalline substances, including vitamin B_{12}, vitamin D, penicillin, and insulin. Her studies showed how the atoms in these materials are organized.

What's X-ray crystallography?

Using X rays, scientists are able to figure out the shapes and the atomic structures of crystalline substances.

As X rays travel through a crystal, the crystal diffracts, or scatters, the X rays into a regular pattern. Like an individual's fingerprints, each crystalline substance has a unique diffraction pattern.

Crystallography has applications in the life, Earth, and physical sciences. For example, geologists use X-ray crystallography to study minerals found in Earth's rocks. Minerals like diamond and graphite have the same chemical makeup but different atomic structures. Because the atomic structure of a mineral is unique, X-ray crystallography helps geologists know exactly which substance they are examining.

Crowfoot Hodgkin

Trailblazing scientist and humanitarian

How does Hodgkin's research help people today?

Dr. Hodgkin's discovery of the structure of insulin helped scientists learn how to control diabetes, a disease that affects over 15 million Americans. Diabetics' bodies are unable to process sugar efficiently. Diabetes can be fatal. Fortunately, Dr. Hodgkin's research with insulin has saved many lives.

What were some obstacles Hodgkin overcame?

Dorothy Hodgkin was born in 1910. When she was four, she and her sisters were left in England while their parents traveled to Egypt.

Dr. Hodgkin always said that this helped encourage her independent spirit. During the 1930s, there were few women scientists. It was a man's world. Hodgkin was not even allowed to attend meetings of the chemistry faculty where she taught because she was a woman. Eventually, she won over her stubborn colleagues with her intelligence and tenacity. Hodgkin died in 1994.

What else did she do?

Dr. Hodgkin was involved in world peace, nuclear disarmament, and aiding poorer nations. She also helped make it a bit easier for women scientists to be accepted by men.

CONNECTIONS Research **Look in reference books or go to the Glencoe Science Web site for information on how X-ray crystallography is used to study minerals. Write your findings and share them with your class.**

SCIENCE *Online*

For more information, visit
science.glencoe.com

Reviewing Main Ideas

Section 1 Minerals

1. You use minerals every day of your life. The pencil "lead" you write with is the mineral graphite. The pretzels you eat for a snack are sprinkled with halite. Gold, silver, gems, and semiprecious stones are used in jewelry. Much of what you use each day is made at least in some part from minerals.

2. All minerals are formed by natural processes and are inorganic solids with definite chemical compositions and orderly internal atomic structures. *Why do minerals in geodes like the one shown here have nicely formed crystal faces?*

3. Minerals have crystal structures in one of six major crystal systems.

Section 2 Mineral Identification

1. Hardness is a measure of how easily a mineral can be scratched.

2. Luster describes how light reflects from a mineral's surface. *What type of luster does this cube-shaped mineral have?*

3. Streak is the color of the powder left by a mineral on an unglazed porcelain tile.

4. Minerals that break along smooth, flat surfaces have cleavage. When minerals break with rough or jagged surfaces, they are displaying fracture.

5. Some minerals have special properties that aid in identifying them. For example, clear crystals of calcite display a double image when placed over type on a page. The mineral magnetite is identified readily by its attraction to a magnet.

Section 3 Uses of Minerals

1. Gems are minerals that are more rare and beautiful than common minerals.

2. Minerals are useful for their physical properties and for the elements they contain. Titanium is a useful element that is derived from the minerals ilmenite and rutile. Rocks containing ilmenite and rutile are ores of titanium, provided that they can be mined at a profit. *What physical property of diamond makes it useful as an industrial tool as well as a beautiful gem?*

FOLDABLES
Reading & Study Skills

After You Read

Exchange Foldables with a classmate and quiz each other to see if you know the answers to your mineral questions.

Visualizing Main Ideas

Complete the following concept map about minerals. Use the following words and phrases: the way a mineral breaks, the way a mineral reflects light, ore, a rare and beautiful mineral, how easily a mineral is scratched, streak, *and* a useful substance mined for profit.

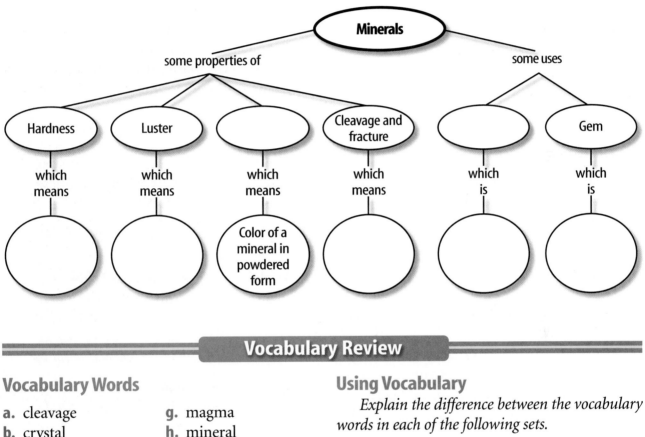

Vocabulary Review

Vocabulary Words

a. cleavage
b. crystal
c. fracture
d. gem
e. hardness
f. luster
g. magma
h. mineral
i. silicate
j. specific gravity
k. streak

Using Vocabulary

Explain the difference between the vocabulary words in each of the following sets.

1. cleavage, fracture
2. crystal, mineral
3. luster, streak
4. magma, crystal
5. hardness, specific gravity
6. magma, mineral
7. crystal, luster
8. mineral, silicate
9. gem, crystal
10. streak, specific gravity

THE PRINCETON REVIEW **Study Tip**

After you read a chapter, write ten questions that it answers. Wait one day and then answer your questions. Look up what you can't remember.

Chapter 1 Assessment

Checking Concepts

Choose the word or phrase that best answers the question.

1. Which is a characteristic of a mineral?
   A) It can be a liquid.
   B) It is organic.
   C) It has no crystal structure.
   D) It is inorganic.

2. What must all silicates contain?
   A) magnesium
   B) silicon and oxygen
   C) silicon and aluminum
   D) oxygen and carbon

3. What is hot, melted rock called?
   A) magma
   B) quartz
   C) salt water
   D) gypsum

4. Which mineral group does quartz belong to?
   A) oxides
   B) carbonates
   C) silicates
   D) sulfides

5. What is the measure of how easily a mineral can be scratched?
   A) luster
   B) hardness
   C) cleavage
   D) fracture

6. What is the color of a powdered mineral formed when rubbing it against an unglazed porcelain tile?
   A) luster
   B) density
   C) hardness
   D) streak

7. In what way does quartz break?
   A) cleavage
   B) fracture
   C) luster
   D) flat planes

8. Which of the following must crystalline solids have?
   A) carbonates
   B) cubic structures
   C) ordered atoms
   D) cleavage

9. Which is hardest on Mohs scale?
   A) talc
   B) quartz
   C) diamond
   D) feldspar

10. Which crystal system do halite crystals belong to?
    A) triclinic
    B) hexagonal
    C) monoclinic
    D) cubic

Thinking Critically

11. Water is a nonliving substance that is formed by natural processes on Earth. It has a unique composition. Sometimes water is a mineral and other times it is not. Explain.

12. How many sides does a perfect salt crystal have?

13. Suppose you let a sugar solution evaporate, leaving sugar crystals behind. Are these crystals minerals? Explain.

14. Will diamond leave a streak on a streak plate? Explain.

15. Explain how you would use **Table 1** to determine the hardness of any mineral.

Developing Skills

16. **Drawing Conclusions** Suppose you found a white mineral with a glassy, nonmetallic luster that was harder than calcite. You identify the sample as quartz. What are your observations? What is your conclusion?

17. **Interpreting Data** Using **Table 1,** identify a mineral with these properties: pink color, nonmetallic and glassy luster, softer than topaz and quartz, scratches apatite, harder than fluorite, has cleavage, and is scratched by a steel file.

18. **Collecting Data** Make an outline of how at least seven physical properties can be used to identify unknown materials.

19. **Measuring in SI** The volume of water in a graduated cylinder is 107.5 mL. A specimen of quartz, tied to a piece of string, is immersed in the water. The new water level reads 186 mL. What is the volume of the piece of quartz?

20. **Concept Mapping** Make a network tree concept map showing two crystal systems and two examples from each group. Use the following words and phrases: *hexagonal, corundum, halite, fluorite,* and *quartz.*

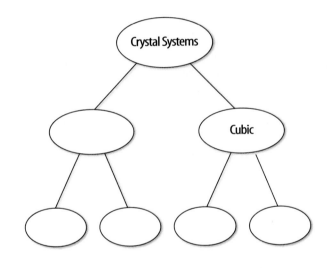

Performance Assessment

21. **Poster** Make a poster that shows the six crystal systems of minerals. Research the crystal systems of minerals and give three examples for each crystal system. Are any of the minerals found in your state? Do any of the minerals have an important use? Display your poster for the class.

TECHNOLOGY

Go to the Glencoe Science Web site at **science.glencoe.com** or use the **Glencoe Science CD-ROM** for additional chapter assessment.

Test Practice

Stacey's science teacher gave her nine different mineral samples that are on the Mohs Hardness Scale (see chart below). Her task was to identify each sample using hardness tests.

Mohs Hardness Scale		Approximate Hardness of Common Objects	
Talc	1	Fingernail	(2.5)
Gypsum	2		
Calcite	3	Copper penny	(3.0)
Fluorite	4		
Apatite	5	Iron nail	(4.5)
Feldspar	6		
Quartz	7	Glass	(5.5)
Topaz	8		
Corundum	9	Steel file	(6.5)
Diamond	10	Streak plate	(7.0)

Study the diagram and answer the following questions.

1. One of Stacey's mineral samples can scratch glass but is scratched by a steel file. According to the chart, which mineral is it?
   A) calcite   **C)** feldspar
   B) apatite   **D)** quartz

2. Which of the following is a method that will distinguish samples of topaz and corundum?
   F) scratching glass with both samples
   G) scratching both samples with a steel file
   H) scratching a streak plate with both samples
   J) scratching the samples on each other

Rocks

Have you ever seen a rock and wondered where it might have come from or how old it is? Whether it is a small pebble found by the side of the road or a mountain of solid rock, every rock is like a history book that tells how and where it was formed. In this chapter, you will learn about how rocks form. You'll learn about the three different groups of rocks—igneous, sedimentary, and metamorphic. You also will learn about the cycle that describes how rocks change from one type to another.

What do you think?

Science Journal Look at the picture below with a classmate. Discuss what you think this might be. Here's a hint: *Your teacher could use this to announce an assignment.* Write your answer or best guess in your Science Journal.

El Capitan rises straight up from the valley floor in Yosemite National Park, California. The hard rock that attracts rock climbers and sightseers from around the world is made of small mineral grains that lock together like pieces of a puzzle. Other rocks form from grains of sand tightly held together or from lava flowing from a volcano. If you examine rocks closely, you sometimes can tell what they are made of.

Safety Precautions 🧤 🥽 🧪

Determine what rocks are made of

1. Collect three or four different rock fragments near your home or school.

2. Draw a picture of the details you see in each rock.

3. Look for different types of materials within the same rock. If these different materials could be separated by physical means, then the rock would be considered a mixture.

Observe

Describe the characteristics of each rock. Are your rocks mixtures? If so, what might these mixtures contain? Write your observations in your Science Journal.

Before You Read

FOLDABLES
Reading & Study Skills

Making an Organizational Study Fold Make the following Foldable to help you organize your thoughts into clear categories about types of rock.

1. Stack two sheets of paper in front of you so the short side of both sheets is at the top.

2. Slide the top sheet up so about four centimeters of the bottom sheet show.

3. Fold both sheets top to bottom to form four tabs and staple along the topfold as shown.

4. Title the top flap *Three Types of Rocks*. Label the other flaps *Igneous, Metamorphic,* and *Sedimentary*.

5. As you read the chapter, record what you learn about the three types of rocks on the flaps.

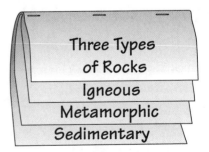

The Rock Cycle

As You Read

What **You'll Learn**

■ **Distinguish** between a rock and a mineral.
■ **Describe** the rock cycle and some changes that a rock could undergo.

Vocabulary
rock
rock cycle

Why **It's Important**

Rocks are everywhere around you—near your school, your home, and even in the sidewalks you walk on.

What is a rock?

Imagine it's the top of the sixth inning. With the sweltering Sun at your back, you scrape your shoe along the infield, waiting for the pitch. Among all the sand and dull-looking, gray stones, your eye catches a glint from a piece of rock that has shiny crystals in it. You quickly snatch it up and put it in your pocket for further examination after the game.

Common Rocks The next time you walk past a large building or monument, stop and take a close look at it. Chances are that it is made out of common rock. In fact, most rock used for building stone contains one or more common minerals, called rock-forming minerals, such as quartz, feldspar, mica, or calcite. When you look closely, the sparkles you see are individual crystals of minerals. A **rock** is a mixture of such minerals, volcanic glass, organic matter, or other materials. **Figure 1** shows minerals mixed together to form the rock granite. You might even find granite on your baseball field.

feldspar

quartz

mica

hornblende

Figure 1
Mount Rushmore, in South Dakota, is made of granite. Granite is a mixture of feldspar, quartz, mica, hornblende, and other minerals.

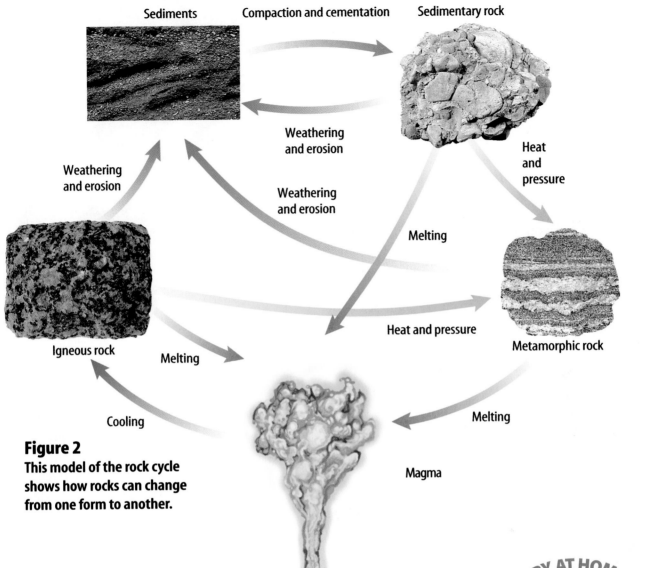

Sediments

Compaction and cementation

Sedimentary rock

Weathering and erosion

Weathering and erosion

Weathering and erosion

Heat and pressure

Weathering and erosion

Melting

Igneous rock

Melting

Heat and pressure

Metamorphic rock

Cooling

Melting

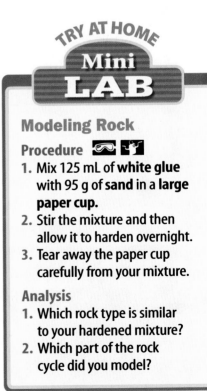

Magma

Figure 2
This model of the rock cycle shows how rocks can change from one form to another.

The Rock Cycle To show how rocks slowly change through time, scientists have created a model called the **rock cycle,** shown in **Figure 2.** It illustrates the processes that create and change rocks. The rock cycle shows the three types of rock—igneous, metamorphic, and sedimentary—and the processes that form them. Look at the rock cycle and notice that rocks change by many processes. For example, a sedimentary rock can change by heat and pressure to form a metamorphic rock. The metamorphic rock then can melt and later cool to form an igneous rock. The igneous rock then could be broken into fragments by weathering and erode away. The fragments might later compact and cement together to form another sedimentary rock. Any given rock can change into any of the three major rock types. A rock even can transform into another rock of the same type. How many paths can you see through the rock cycle?

✔ **Reading Check** *What is illustrated by the rock cycle?*

Modeling Rock

Procedure 🥽 👕
1. Mix 125 mL of **white glue** with 95 g of **sand** in a **large paper cup.**
2. Stir the mixture and then allow it to harden overnight.
3. Tear away the paper cup carefully from your mixture.

Analysis
1. Which rock type is similar to your hardened mixture?
2. Which part of the rock cycle did you model?

Figure 3

Rocks continuously form and transform in a process that geologists call the rock cycle. For example, molten rock—from volcanoes such as Washington's Mount Rainier, background—cools and solidifies to form igneous rock. It slowly breaks down when exposed to air and water to form sediments. These sediments are compacted or cemented into sedimentary rock. Heat and pressure might transform sedimentary rock into metamorphic rock. When metamorphic rock melts and hardens, igneous rock forms again. There is no distinct beginning, nor is there an end, to the rock cycle.

▲ The black sand beach of this Polynesian island is sediment weathered and eroded from the igneous rock of a volcano nearby.

▲ This alluvial fan on the edge of Death Valley, California, was formed when gravel, sand, and finer sediments were deposited

▲ Layers of shale and chalk form Kansas's Monument Rocks. They are remnants of sediments deposited on

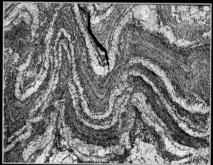

▲ Heat and pressure deep below Earth's surface can change rock into metamorphic rock, like this banded gneiss.

Physics
INTEGRATION

Matter and the Rock Cycle The rock cycle, illustrated in **Figure 3**, shows how rock can be weathered to small rock and mineral grains. This material then can be eroded and carried away by wind, water, or ice. When you think of erosion, it might seem that the material is somehow destroyed and lost from the cycle, but this is not the case. The chemical elements that make up minerals and rocks are not destroyed. This fact illustrates the principle of conservation of matter. The changes that take place in the rock cycle never destroy or create matter. The elements are just redistributed in other forms.

✔ **Reading Check** *What is the principle of conservation of matter?*

Discovering the Rock Cycle James Hutton, a Scottish physician and naturalist, first recognized in 1788 that rocks undergo profound changes. Hutton noticed, among other things, that some layers of solid rock in Siccar Point, shown in **Figure 4,** had been altered since they formed. Instead of showing a continuous pattern of horizontal layering, some of the rock layers at Siccar Point are tilted and partly eroded. However, the younger rocks above them are nearly horizontal.

Hutton published these and other observations, which proved that rocks are subject to constant change. Hutton's early recognition of the rock cycle continues to influence geologists—even after over 200 years have passed.

Figure 4
The rock formations at Siccar Point, Scotland show that rocks undergo constant change.

Section 1 Assessment

1. What materials are rocks made of?
2. What are the three basic types of rock?
3. Describe the major processes of the rock cycle.
4. Describe one way that the rock cycle can illustrate the principle of conservation of matter.
5. **Think Critically** Look at the model of the rock cycle. How would you define magma based on the illustration in **Figure 2?** How would you define sediment and sedimentary rock?

Skill Builder Activities

6. **Concept Mapping** Using a computer, make a concept map explaining how igneous rocks can become sedimentary, then metamorphic, and finally, other igneous rocks. **For more help, refer to the** Science Skill Handbook.

7. **Communicating** Review the model of the rock cycle in **Figure 2.** In your Science Journal, write a story or poem that explains what can happen to a sedimentary rock as it changes through the rock cycle. **For more help, refer to the** Science Skill Handbook.

Igneous Rocks

As You Read

What **You'll Learn**

- **Recognize** magma and lava as the materials that cool to form igneous rocks.
- **Contrast** the formation of intrusive and extrusive igneous rocks.
- **Contrast** granitic and basaltic igneous rocks.

Vocabulary

igneous rock	extrusive
lava	basaltic
intrusive	granitic

Why **It's Important**

Igneous rocks are the most abundant kind of rock in Earth's crust. They contain many valuable resources.

Formation of Igneous Rocks

Perhaps you've heard of recent volcanic eruptions in the news. When some volcanoes erupt, they eject a flow of molten rock material, as shown in **Figure 5.** Molten rock material, called magma, flows when it is hot and becomes solid when it cools. When hot magma cools and hardens, it forms **igneous** (IHG nee us) **rocks.** Why do volcanoes erupt, and where does the molten material come from?

Magma In certain places within Earth, the temperature and pressure are just right for rocks to melt and form magma. Most magmas come from deep below Earth's surface. Magma is located at depths ranging from near the surface to about 150 km below the surface. Temperatures of magmas range from about 650°C to 1,200°C, depending on their chemical compositions and pressures exerted on them.

The heat that melts rocks comes from sources within Earth's interior. One source is the decay of radioactive elements within Earth. Some heat is left over from the formation of the planet, which originally was molten. Radioactive decay of elements contained in rocks balances some heat loss as Earth continues to cool.

Because magma is less dense than surrounding solid rock, it is forced upward toward the surface, as shown in **Figure 6.** When magma reaches Earth's surface and flows from volcanoes, it is called **lava.**

Figure 5
Some lava is highly fluid and free-flowing, as shown by this spectacular lava fall in Volcano National Park, East Rift, Kilauea, Hawaii.

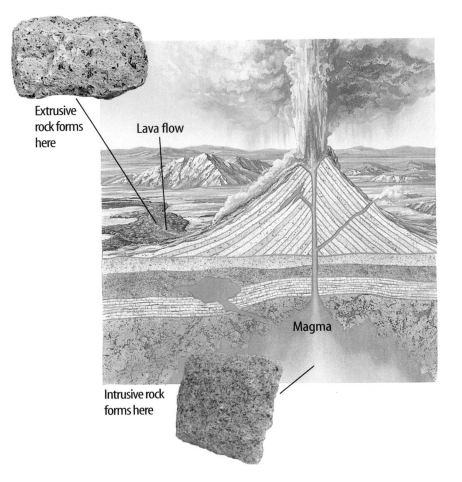

Extrusive rock forms here

Lava flow

Magma

Intrusive rock forms here

Figure 6
Intrusive rocks form from magma trapped below Earth's surface. Extrusive rocks form from lava flowing at the surface.

Are there any igneous rocks used as building stones in your area? To find out more about rocks used in construction, see the **Building Stones Field Guide** at the back of the book.

Intrusive Rocks Magma is made up of atoms and molecules of melted minerals. As magma cools, the atoms and molecules rearrange themselves into new crystals called mineral grains. Rocks form as these mineral grains grow together. Rocks that form from magma below the surface, as illustrated in **Figure 6,** are called **intrusive** igneous rocks. Intrusive rocks are found at the surface only after the layers of rock and soil that once covered them have been removed by erosion. Erosion occurs when the rocks are pushed up by forces within Earth. Because intrusive rocks form at depth and they are surrounded by other rocks, it takes a long time for them to cool. Slowly cooled magma produces individual mineral grains that are large enough to be observed with the unaided eye.

Extrusive Rocks **Extrusive** igneous rocks are formed as lava cools on the surface of Earth. When lava flows on the surface, as illustrated in **Figure 6,** it is exposed to air and water. Lava, such as the basaltic lava shown in **Figure 5,** cools quickly under these conditions. The quick cooling rate keeps mineral grains from growing large, because the atoms and molecules don't have the time to arrange into large crystals. Therefore, extrusive igneous rocks are fine grained.

Reading Check *What controls the grain size of an igneous rock?*

Table 1 Common Igneous Rocks

Magma Type	Basaltic		Andesitic	Granitic		
Intrusive	Gabbro		Diorite	Granite		
Extrusive	Basalt	Scoria	Andesite	Rhyolite	Pumice	Obsidian

SCIENCE
Online

Research Visit the Glencoe Science Web site at **science.glencoe.com** for more information about intrusive and extrusive rocks. Summarize new information you learn in your Science Journal.

Volcanic Glass Pumice, obsidian, and scoria are examples of volcanic glass. These rocks cooled so quickly that few or no mineral grains formed. Most of the atoms in these rocks are not arranged in orderly patterns, and few crystals are present.

In the case of pumice and scoria, gases become trapped in the gooey molten material as it cools. Some of these gases eventually escape, but holes are left behind where the rock formed around the pockets of gas.

Classifying Igneous Rocks

Igneous rocks are intrusive or extrusive depending on how they are formed. A way to further classify these rocks is by the magma from which they form. As shown in **Table 1,** an igneous rock can form from basaltic, andesitic, or granitic magma. The type of magma that cools to form an igneous rock determines important chemical and physical properties of that rock. These include mineral composition, density, color, and melting temperature.

✔ **Reading Check** *Name two ways igneous rocks are classified.*

Basaltic Rocks Basaltic (buh SAWL tihk) igneous rocks are dense, dark-colored rocks. They form from magma that is rich in iron and magnesium and poor in silica, which is the compound SiO_2. The presence of iron and magnesium in minerals in basalt gives basalt its dark color. Basaltic lava is fluid and flows freely from volcanoes in Hawaii, such as Kilauea. How does this explain the black beach sand common in Hawaii?

Granitic Rocks Granitic igneous rocks are light-colored rocks of a lower density than basaltic rocks. Granitic magma is thick and stiff and contains lots of silica but lesser amounts of iron and magnesium. Because granitic magma is stiff, it can build up a great deal of gas pressure, which is released explosively during violent volcanic eruptions.

Andesitic Rocks Andesitic igneous rocks have mineral compositions between those of granitic and basaltic rocks. Many volcanoes around the rim of the Pacific Ocean formed from intermediate magmas. Like volcanoes that erupt granitic magma, these volcanoes also can erupt violently.

Take another look at **Table 1.** Basalt forms at the surface of Earth because it is an extrusive rock. Granite forms below Earth's surface from magma with a high concentration of silica. When you identify an igneous rock, you can infer how it formed and the type of magma that it formed from.

Chemistry INTEGRATION

Inside Earth, materials contained in rocks can melt. In your Science Journal, describe what is happening to the atoms and molecules to cause this change of state.

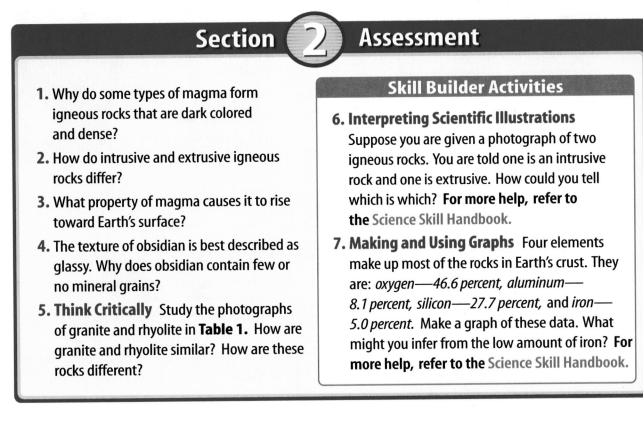

Section 2 Assessment

1. Why do some types of magma form igneous rocks that are dark colored and dense?

2. How do intrusive and extrusive igneous rocks differ?

3. What property of magma causes it to rise toward Earth's surface?

4. The texture of obsidian is best described as glassy. Why does obsidian contain few or no mineral grains?

5. **Think Critically** Study the photographs of granite and rhyolite in **Table 1.** How are granite and rhyolite similar? How are these rocks different?

Skill Builder Activities

6. **Interpreting Scientific Illustrations** Suppose you are given a photograph of two igneous rocks. You are told one is an intrusive rock and one is extrusive. How could you tell which is which? **For more help, refer to the** Science Skill Handbook.

7. **Making and Using Graphs** Four elements make up most of the rocks in Earth's crust. They are: *oxygen—46.6 percent, aluminum— 8.1 percent, silicon—27.7 percent,* and *iron— 5.0 percent.* Make a graph of these data. What might you infer from the low amount of iron? **For more help, refer to the** Science Skill Handbook.

Activity

Igneous Rock Clues

You've learned how color often is used to estimate the composition of an igneous rock. The texture of an igneous rock describes its overall appearance, including mineral grain sizes and the presence or absence of bubble holes, for example. In most cases, grain size relates to how quickly the magma or lava cooled. Crystals you can see without a magnifying glass indicate slower cooling. Smaller, fine-grained crystals indicate quicker cooling, possibly due to volcanic activity. Rocks with glassy textures cooled so quickly that there was no time to form mineral grains.

What You'll Investigate

What does an igneous rock's texture and color indicate about its formation history?

Materials

rhyolite	granite
basalt	obsidian
vesicular basalt	gabbro
pumice	magnifying glass

Goals

- **Classify** different samples of igneous rocks by color and infer their composition.
- **Observe** the textures of igneous rocks and infer how they formed.

Safety Precautions

Some rock samples might have sharp edges. Always use caution while handling samples.

Procedure

1. **Arrange** rocks according to color (light or dark). Record your observations in your Science Journal.
2. **Arrange** rocks according to similar texture. Consider grain sizes and shapes, presence of holes, etc. Use your magnifying glass to see small features more clearly. Record your observations.

Conclude and Apply

1. Infer which rocks are granitic based on color.
2. Infer which rocks cooled quickly. What observations led you to this inference?
3. Identify any samples that suggest gases were escaping from them as they cooled.
4. Which samples have a glassy appearance? How did these rocks form?
5. **Infer** which samples are not volcanic. Explain your choices.

Communicating Your Data

Research the compositions of each of your samples. Did the colors of any samples lead you to infer the wrong compositions? Communicate to your class what you learned.

Metamorphic Rocks

Formation of Metamorphic Rocks

Have you ever packed your lunch in the morning and not been able to recognize it at lunchtime? You might have packed a sandwich, banana, and a large bottle of water. You know you didn't smash your lunch on the way to school. However, you didn't think about how the heavy water bottle would damage your food if the bottle was allowed to rest on the food all day. The heat in your locker and the pressure from the heavy water bottle have changed the form of your sandwich. Like your lunch, rocks can be affected by temperature changes and pressure.

Metamorphic Rocks Rocks that have changed because of changes in temperature and pressure or the presence of hot, watery fluids are called **metamorphic rocks.** Changes that occur can be in the form of the rock, shown in **Figure 7,** the composition of the rock, or both. Metamorphic rocks can form from igneous, sedimentary, or other metamorphic rocks. What Earth processes can change these rocks?

As You Read

What **You'll Learn**

■ **Describe** the conditions in Earth that cause metamorphic rocks to form.
■ **Classify** metamorphic rocks as foliated or nonfoliated.

Vocabulary

metamorphic rock
foliated
nonfoliated

Why **It's Important**

Metamorphic rocks are useful because of their unique properties.

+ pressure

Figure 7

A The mineral grains in granite are flattened and aligned when heat and pressure are applied to them. **B** As a result, gneiss is formed. *What other conditions can cause metamorphic rocks to form?*

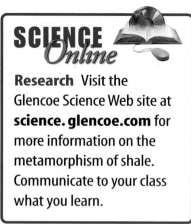
Heat and Pressure Rocks beneath Earth's surface are under great pressure from rock layers above them. Temperature also increases with depth in Earth. In some places, the heat and pressure are just right to cause rocks to melt and magma to form. In other areas where melting doesn't occur, some mineral grains can become flattened like the sandwich in the lunch bag. Sometimes, under these conditions, minerals exchange atoms with surrounding minerals and new, bigger minerals form.

Depending upon the amount of pressure and temperature applied, one type of rock can change into several different metamorphic rocks, and each type of metamorphic rock can come from several kinds of parent rocks. For example, the sedimentary rock shale will change into slate. As increasing pressure and temperature are applied, the slate can change into phyllite, then schist, and eventually gneiss. Schist also can form when basalt is metamorphosed, or changed, and gneiss can come from granite.

✔ **Reading Check** *How can one type of rock change into several different metamorphic rocks?*

Hot Fluids Did you know that fluids can move through solid rock? These fluids, which are mostly water with dissolved elements, can react chemically with a rock and change its composition, especially when the fluids are hot. That's what happens when rock surrounding a hot magma body reacts with hot fluids from the magma, as shown in **Figure 8.** Most fluids that transform rocks during metamorphic processes are hot and mainly are comprised of water and carbon dioxide.

Figure 8
In the presence of hot, water-rich fluids, solid rock can change in mineral composition without having to melt.

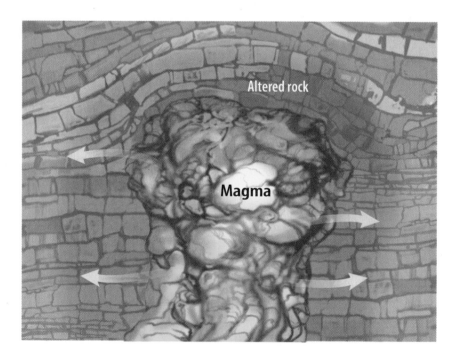

Altered rock

Magma

Classifying Metamorphic Rocks

Metamorphic rocks form from igneous, sedimentary, or other metamorphic rocks. Heat, pressure, and hot fluids trigger the changes. Each resulting rock can be classified according to its composition and texture.

Foliated Rocks When mineral grains flatten and line up in parallel layers, the metamorphic rock is said to have a **foliated** or banded texture. Two examples of foliated rocks are slate and gneiss. Slate forms from the sedimentary rock shale. The minerals in shale arrange into layers when they are exposed to heat and pressure. As **Figure 9** shows, slate separates easily along these foliation layers.

The minerals in slate are pressed together so tightly that water can't pass between them easily. Because it's watertight, slate is ideal for paving around pools and patios. The naturally flat nature of slate and the fact that it splits easily make it useful for roofing and tiling many surfaces.

Gneiss (NISE), another foliated rock, forms when granite and other rocks are changed. Quartz, feldspar, mica (MI kuh), and other minerals that make up granite aren't changed much, but they arrange into alternating bands of light and dark minerals.

Figure 9
Slate often is used as a building or landscaping material. *What property makes slate so useful for these purposes?*

✔ **Reading Check** *What type of metamorphic rock is composed of layers of mineral grains?*

Figure 10
This exhibit in Vermont shows the beauty of carved marble.

Nonfoliated Rocks In some metamorphic rocks, layering does not occur. The mineral grains grow and rearrange, but they don't form layers. This process produces a **nonfoliated** texture.

Sandstone is a sedimentary rock that's often composed mostly of quartz grains. When sandstone is heated under a lot of pressure, the grains of quartz grow in size and become interlocking, like the pieces of a jigsaw puzzle. The resulting rock is called quartzite.

Marble is another nonfoliated metamorphic rock. Marble forms from the sedimentary rock limestone, which is composed of the mineral calcite. Usually, marble contains several other minerals besides calcite. For example, hornblende and serpentine give marble a black or greenish tone, whereas hematite makes it red. As **Figure 10** shows, marble is a popular material for artists to sculpt because it is not as hard as other rocks.

So far, you've investigated only a portion of the rock cycle. You still haven't observed how sedimentary rocks are formed and how igneous and metamorphic rocks evolve from them. The next section will complete your investigation of the rock cycle.

Section 3 Assessment

1. How is the formation of metamorphic rocks different from that of igneous, or sedimentary, rocks?

2. What role do fluids play in rock metamorphism?

3. How are metamorphic rocks classified? What are the characteristics of rocks in each of these classifications?

4. Give an example of a foliated and a nonfoliated metamorphic rock. Name one of their possible parent rocks.

5. **Think Critically** Marble is a common material used to make sculptures, but not just because it's a beautiful stone. What properties of marble make it useful for this purpose?

Skill Builder Activities

6. **Concept Mapping** Put the following events in an events-chain concept map that explains how a metamorphic rock might form from an igneous rock. Here's a hint: *Start with "igneous rock forms."* Use each event just once. **For more help, refer to the** Science Skill Handbook.

   Events: *sedimentary rock forms, weathering occurs, heat and pressure are applied, igneous rock forms, metamorphic rock forms, erosion occurs, sediments are formed, deposition occurs*

7. **Using Graphics Software** Use a graphics program to illustrate how metamorphic rocks form. Be sure to show how directed pressure causes alignment of mineral grains. **For more help, refer to the** Technology Skill Handbook.

4 Sedimentary Rocks

Formation of Sedimentary Rocks

Igneous rocks are the most common rocks on Earth, but because most of them exist below the surface, you might not have seen too many of them. That's because 75 percent of the rocks exposed at the surface are sedimentary rocks.

Sediments are loose materials such as rock fragments, mineral grains, and bits of shell that have been moved by wind, water, ice, or gravity. If you look at the model of the rock cycle, you will see that sediments come from already-existing rocks that are weathered and eroded. **Sedimentary rocks** form when sediments are pressed and cemented together, or when minerals form from solutions.

Stacked Rocks Sedimentary rocks often form as layers. The older layers are on the bottom because they were deposited first. Sedimentary rock layers are a lot like the books and papers in your locker. Last week's homework is on the bottom, and today's notes will be deposited on top of the stack. However, if you disturb the stack, the order in which the books and papers are stacked will change, as shown in **Figure 11.** Sometimes, forces within Earth overturn layers of rock, and the oldest are no longer on the bottom.

As You Read

What **You'll Learn**
- **Explain** how sedimentary rocks form from sediments.
- **Classify** sedimentary rocks as detrital, chemical, or organic in origin.

Vocabulary

sediment	compaction
sedimentary rock	cementation

Why **It's Important**
Some sedimentary rocks, like coal, are important sources of energy.

Figure 11
Like sedimentary rock layers, the oldest paper is at the bottom of the stack. If the stack is disturbed, then it is no longer in order.

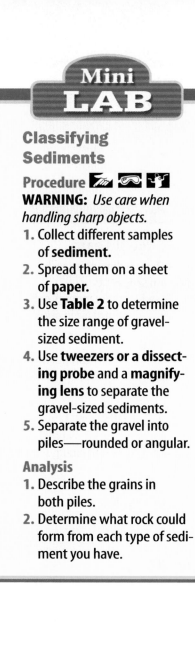
Figure 12
During compaction, pore space between sediments decreases, causing them to become packed together more tightly.

Classifying Sedimentary Rocks

Sedimentary rocks can be made of just about any material found in nature. Sediments come from weathered and eroded igneous, metamorphic, and sedimentary rocks. Sediments also come from the remains of some organisms. The composition of a sedimentary rock depends upon the composition of the sediments from which it formed.

Like igneous and metamorphic rocks, sedimentary rocks are classified by their composition and by the manner in which they formed. Sedimentary rocks usually are classified as detrital, chemical, or organic.

Detrital Sedimentary Rocks

The word *detrital* (Dih TRY tul) comes from the Latin word *detritus*, which means "to wear away." Detrital sedimentary rocks, such as those shown in **Table 2,** are made from the broken fragments of other rocks. These loose sediments are compacted and cemented together to form solid rock.

Weathering and Erosion When rock is exposed to air, water, or ice, it is unstable and breaks down chemically and mechanically. This process, which breaks rocks into smaller pieces, is called weathering. **Table 2** shows how these pieces are classified by size. The movement of weathered material is called erosion.

Compaction Erosion moves sediments to a new location, where they then are deposited. Here, layer upon layer of sediment builds up. Pressure from the upper layers pushes down on the lower layers. If the sediments are small, they can stick together and form solid rock. This process, shown in **Figure 12,** is called **compaction.**

✔ **Reading Check** *How do rocks form through compaction?*

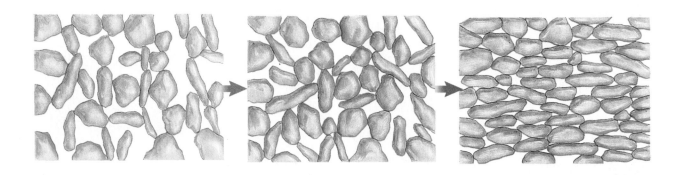

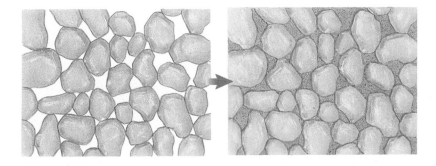

Figure 13
Sediments are cemented together as minerals crystallize between grains.

Cementation If sediments are large, like sand and pebbles, pressure alone can't make them stick together. Large sediments have to be cemented together. **Cementation,** which is shown in **Figure 13,** occurs when water soaks through soil and rock. As water moves through soil and rock, it picks up atoms and molecules released from minerals during weathering. The resulting solution of water and dissolved materials moves through open spaces between sediments. Minerals such as quartz, calcite, hematite, and limonite are deposited between the pieces of sediment. These minerals, acting as natural cements, hold the sediment together like glue, making a detrital sedimentary rock.

Shape and Size of Sediments Detrital rocks have granular textures much like granulated sugar. They are named according to the shapes and sizes of the sediments that form them. For example, conglomerate and breccia both form from large sediments, as shown in **Table 2.** If the sediments are rounded, the rock is called conglomerate. If the sediments have sharp angles, the rock is called breccia. The roundness of sediment particles depends on how far they have been moved by wind or water.

Table 2 Sediment Sizes and Detrital Rocks

Sediment	Clay	Silt	Sand	Gravel
Size Range	<0.004 mm	0.004–0.063 mm	0.063–2 mm	>2 mm
Example	Shale	Siltstone	Sandstone	Conglomerate (shown) or Breccia

Conglomerate

Figure 14
Although concrete strongly resembles conglomerate, concrete is not a rock because it does not occur in nature.

Chemistry
INTEGRATION

You may know of large caves or caverns that have formed underground in limestone deposits. Research the chemical reactions that form caves in Carlsbad Caverns, New Mexico, or Mammoth Cave, Kentucky.

Materials Found in Sedimentary Rocks The gravel-sized sediments in conglomerate and breccia can consist of any type of rock or mineral. Often, they are composed of chunks of the minerals quartz and feldspar. They also can be pieces of rocks such as gneiss, granite, or limestone. The cement that holds the sediments together usually is made of quartz or calcite.

Have you ever looked at the concrete in sidewalks, driveways, and stepping stones? The concrete in **Figure 14** is made of gravel and sand grains that have been cemented together. Although the structure is similar to that of naturally occurring conglomerate, it cannot be considered a rock.

Sandstone is formed from smaller particles than conglomerates and breccias. Its sand-sized sediments can be just about any mineral, but they are usually grains of the minerals quartz and feldspar that are compacted and cemented together. Siltstone is similar to sandstone except it is made of smaller, silt-sized particles. Shale is a detrital sedimentary rock that is made mainly of clay-sized particles. Clay-sized sediments are compacted together by pressure from overlying layers.

Chemical Sedimentary Rocks

Chemical sedimentary rocks form when dissolved minerals come out of solution. You can show that salt is deposited in the bottom of a glass or pan when saltwater solution evaporates. In a similar way, minerals collect when seas or lakes evaporate. The deposits of minerals that come out of solution form sediments and rocks. For example, the sediment making up New Mexico's White Sands desert consists of pieces of a chemical sedimentary rock called rock gypsum. Chemical sedimentary rocks are different. They are not made from pieces of preexisting rocks.

Reading Check *How do chemical sedimentary rocks form?*

Limestone Calcium carbonate is carried in solution in ocean water. When calcium carbonate comes out of solution as calcite and its many crystals grow together, limestone forms. Limestone also can contain other minerals and sediments, but it must be at least 50 percent calcite. Limestone usually is deposited on the bottom of lakes or shallow seas. Large areas of the central United States have limestone bedrock because seas covered much of the country for millions of years. It is hard to imagine Kansas being covered by ocean water, but it has happened several times throughout geological history.

Rock Salt When water that is rich in dissolved salt evaporates, it often deposits the mineral halite. Halite forms rock salt, shown in **Figure 15.** Rock salt deposits can range in thickness from a few meters to more than 400 m. Companies mine these deposits because rock salt is an important resource. It's used in the manufacturing of glass, paper, soap, and dairy products. The halite in rock salt is used as table salt.

Figure 15
Rock salt is extracted from this mine in Germany. The same salt can be processed and used to season your favorite foods.

Organic Sedimentary Rocks

Rocks made of the remains of once-living things are called organic sedimentary rocks. One of the most common organic sedimentary rocks is fossil-rich limestone. Like chemical limestone, fossil-rich limestone is made of the mineral calcite ($CaCO_3$). However, fossil-rich limestone mostly contains remains of once-living ocean organisms instead of only calcite that has come out of ocean water.

Animals such as mussels, clams, corals, and snails make their shells from $CaCO_3$ that eventually becomes calcite. When they die, their shells accumulate on the ocean floor. When these shells are cemented together, fossil-rich limestone forms. If a rock is made completely of shell fragments that you can see, the rock is called coquina (koh KEE nuh).

Chalk Chalk is another organic sedimentary rock that is made of microscopic shells. When you write with naturally occurring chalk, you're crushing and smearing the calcite-shell remains of once-living ocean organisms.

Coal Another useful organic sedimentary rock is coal, shown in **Figure 16.** Coal forms when pieces of dead plants are buried under other sediments in swamps. These plant materials are chemically changed by microorganisms. The resulting sediments are compacted over millions of years to form coal, an important source of energy. Much of the coal in North America and Europe formed during a period of geologic time that is so named because of this important reason. The Carboniferous Period, which spans from approximately 360 to 286 million years ago, was named in Europe. So much coal formed during this interval of time that coal's composition—primarily carbon—was the basis for naming a geologic period.

Math Skills Activity

Calculating Thickness

Example Problem

It took 300 million years for a layer of plant matter about 0.9m thick to produce a bed of bituminous coal 0.3m thick. Estimate the thickness of plant matter that produced a bed of coal 0.15m thick.

Solution

1 *This is what you know:*

original thickness of plant matter = 0.9m
original coal thickness = 0.3m
new coal thickness = 0.15m

2 *This is what you need to know:*

thickness of plant matter needed to form 0.15m of coal

3 *This is the equation you need to use:*

(thickness of plant matter) / (new coal thickness) = (original thickness of plant matter) / original coal thickness

4 *Substitute the known values:*

(?m plant matter) / (0.15m coal) = (0.9m plant matter) / (0.3m coal)

5 *Solve the equation:*

(?m plant matter) = (0.9m plant matter) (0.15m coal) / (0.3m coal) = 0.45m plant matter

Multiply your answer by the original coal thickness. Divide by original plant matter thickness. Do you get the new coal thickness that was given?

Practice Problem

Estimate the thickness of plant matter that produced a bed of coal 0.6m thick.

For more help, refer to the Math Skill Handbook.

Figure 16
This coal layer in Alaska is easily identified by its jet-black color, as compared with other sedimentary layers.

Another Look at the Rock Cycle

You have seen that the rock cycle has no beginning and no end. Rocks change continually from one form to another. Sediments come from rocks and minerals that have been broken apart. Even the magma that forms igneous rocks comes from the melting of rocks that already exist.

All of the rocks that you've learned about in this chapter formed through some process within the rock cycle. All of the rocks around you, including those used to build houses and monuments, are part of the rock cycle. Slowly, they are all changing, because the rock cycle is a continuous, dynamic process.

Section 4 Assessment

1. Where do sediments come from?

2. Explain how compaction and cementation are important in forming sedimentary rocks.

3. Explain the difference between detrital and chemical sedimentary rock.

4. List chemical sedimentary rocks that are essential to your health or that are used to make life more convenient. How is each used?

5. **Think Critically** Use the rock cycle to explain how pieces of granite and slate could be found in the same piece of conglomerate.

Skill Builder Activities

6. **Making and Using Tables** You are told to identify several rocks based on their compositions and other physical properties. What physical properties would you consider? How would you use these properties so the rocks could be identified easily? **For more help, refer to the** Science Skill Handbook.

7. **Calculating Ratios** Sediment sizes of different sedimentary rock types are presented in **Table 2.** Estimate how many times larger the largest grains of silt and sand are compared to clay grains. **For more help, refer to the** Math Skill Handbook.

Activity

Sedimentary Rocks

S edimentary rocks are formed by compaction and cementation of sediment. Because sediment is found in all shapes and sizes, do you think these characteristics could be used to classify detrital sedimentary rocks? Sedimentary rocks also can be classified as chemical or organic.

What You'll Investigate

How are rock characteristics used to classify sedimentary rocks as detrital, chemical, or organic?

Goals
- **Observe** sedimentary rock characteristics.
- **Compare and contrast** sedimentary rock textures.
- **Classify** sedimentary rocks as detrital, chemical, or organic.

Materials
unknown sedimentary rock samples
marking pen
5 percent hydrochloric acid (HCl)
  vinegar
  dropper
  paper towels
  water
  magnifying lens
  metric ruler
  Alternate materials

Safety Precautions

Sedimentary Rock Samples					
Sample	Observations	Minerals or Fossils Present	Sediment Size	Detrital, Chemical, or Organic	Rock Name
A					
B					
C					
D					
E					

Procedure

1. Make a Sedimentary Rock Samples chart similar to the one shown above in your Science Journal.

2. **Determine** the sizes of sediments in each sample, using a magnifying lens and a metric ruler. Using **Table 2**, classify any grains of sediment in the rocks as gravel, sand, silt, or clay. In general, the sediment is silt if it is gritty and just barely visible, and clay if it is smooth and if individual grains are not visible.

3. Place a few drops of HCl or vinegar on each rock sample. Bubbling on a rock

indicates the presence of calcite. **WARNING:** *HCl is an acid and can cause burns. Wear goggles and a lab apron. Rinse spills with water and wash hands afterwards.*

4. **Examine** each sample for fossils and describe any that are present.

5. **Determine** whether each sample has a granular or nongranular texture.

6. **Classify** your samples as detrital, chemical, or organic. Identify each rock sample.

Conclude and Apply

1. Explain why you tested the rocks with acid. What minerals react with acid?

2. **Compare and contrast** sedimentary rocks that have a granular texture with sedimentary rocks that have a nongranular texture.

Communicating Your Data

Compare your conclusions with those of other students in your class. **For more help, refer to the** Science Skill Handbook.

Australia's controversial rock star

One of the most famous rocks in the world is causing serious problems for Australians

Uluru (yew LEW rew) is one of the most popular tourist destinations in Australia. This sandstone sky-scraper is more than 8 km around, over 300 m high, and extends as much as 4.8 km below the surface. One writer describes it as an iceberg in the desert. Geologists hypothe-size that the mighty Uluru rock began forming 550 million years ago during Precambrian time. That's when large mountain ranges started to form in Central Australia. About 300 million years later, sediment from these ranges settled on the Amadeus Basin. The sandy material hardened into rock. Faults in Earth weakened the rock. Erosion eventually cut away this weakened surrounding rock to expose the massive sandstone structure called Ayers Rock. Today, erosion continues to slowly cut away at the rock.

For more than 25,000 years, this geo-logical wonder has played an important role in the lives of the Aboriginal peoples, the Anangu (a NA noo). These native Australians are the original owners of the rock and have spiritual explanations for its many caves, holes, and scars.

In 1873, the rock was named in honor of Sir Henry Ayers, an Australian government official. Few people in the early 1900s visited it. But as transportation technology improved, so did the rock's tourist appeal. In 1958, the Australian government made Ayers Rock part of a national park. In the 1980s, 100,000 tourists visited it. In 2000, the rock attracted about 400,000 tourists.

North America

Europe

Asia

Africa

Uluru

Australia

Australia

Tourists Take Over

This increase in tourism is causing serious problems. The Anangu take offense at the climbing of their sacred rock, a holy site. However, if climbing the rock were outlawed, tourism would be seriously hurt. And that would mean less income for Australians. One Australian marketing director sums up the apparent threat to tourism: Going to Uluru and not climbing is like going to the Great Barrier Reef and not diving.

Some steps have been taken to respect the native people's wishes. In 1985, the government returned Ayers rock to the Anangu, and agreed to call it by its traditional name. As part of this arrangement, the Anangu leased back the rock to the Australian government until the year 2084, when management of Uluru and the surrounding park goes back to the Anangu. Until then, the Anangu will collect 25 percent of the money people pay to visit the rock.

The Aboriginal people encourage tourists to respect their beliefs. They offer a walking tour around the rock, and they show videos about Aboriginal traditions. The Anangu sell T-shirts that say "I *didn't* climb Uluru." They hope visitors to Uluru will wear the T-shirt with pride and respect.

Athlete Nova Benis-Kneebone had the honor of receiving the Olympic torch near the sacred Uluru and carried it partway to the Olympic stadium.

CONNECTIONS Write Research a natural landmark or large natural land or water formation in your area. What is the geology behind it? When was it formed? How was it formed? Write a folktale that explains its formation. Share your folktale with the class.

SCIENCE *Online*

For more information, visit science.glencoe.com

Reviewing Main Ideas

Section 1 The Rock Cycle

1. A rock is a mixture of one or more minerals, rock fragments, organic matter, or volcanic glass.

2. The rock cycle includes all processes by which rocks form. *How could an igneous rock change to a metamorphic rock?*

Section 2 Igneous Rocks

1. Magma and lava are molten materials that harden to form igneous rocks.

2. Intrusive igneous rocks form when magma cools slowly below the surface. Extrusive igneous rocks form when lava cools rapidly at the surface.

3. Basalt is dense and dark colored. Granite is light colored and less dense than basalt. Andesite has intermediate density and color, somewhere between basalt and granite. *What kind of rock is shown below?*

Section 3 Metamorphic Rocks

1. Heat, pressure, and fluids can cause metamorphic rocks to form. Metamorphic rocks can have foliated textures or nonfoliated textures.

2. Slate and gneiss are examples of foliated metamorphic rocks. Quartzite and marble are examples of nonfoliated metamorphic rocks. *How are these rocks different?*

Section 4 Sedimentary Rocks

1. Detrital sedimentary rocks form when fragments of rocks and minerals are compacted and cemented together. Detrital rocks always have a granular texture.

2. Chemical sedimentary rocks come out of solution or are left behind by evaporation. Chemical rocks generally have a non-granular texture.

3. Organic sedimentary rocks are made mostly of the remains of once-living organisms. *What kinds of remains make up the rock to the right?*

FOLDABLES
Reading & Study Skills

After You Read

Using the information in your Foldable for help, write examples of each type of rock under the flaps of your Foldable.

Visualizing Main Ideas

Complete the following concept map on rocks. Use the following terms:
organic, metamorphic, foliated, extrusive, igneous, *and* chemical.

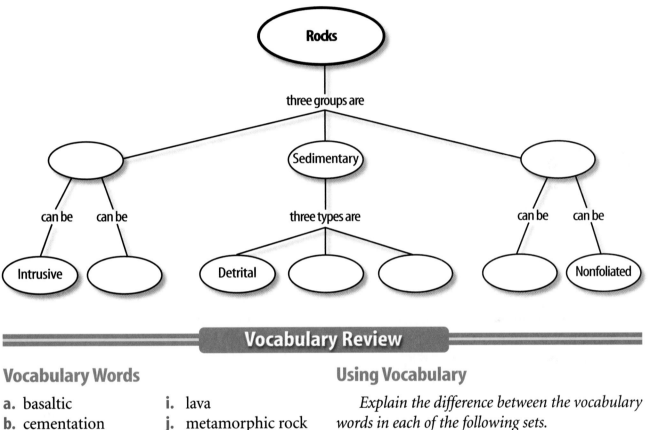

Vocabulary Review

Vocabulary Words

a. basaltic
b. cementation
c. compaction
d. extrusive
e. foliated
f. granitic
g. igneous rock
h. intrusive
i. lava
j. metamorphic rock
k. nonfoliated
l. rock
m. rock cycle
n. sediment
o. sedimentary rock

Using Vocabulary

Explain the difference between the vocabulary words in each of the following sets.

1. foliated, nonfoliated

2. cementation, compaction

3. sediment, lava

4. extrusive, intrusive

5. rock, rock cycle

6. metamorphic rock, igneous rock, sedimentary rock

7. sediment, sedimentary rock

8. lava, igneous rock

9. rock, sediment

10. basaltic, granitic

THE PRINCETON REVIEW **Study Tip**

If you're not sure of the relationship between terms in a question, try making a concept map of the terms to see how they fit together. Ask your teacher if the relationships you drew are correct.

Chapter ② Assessment

Checking Concepts

Choose the word or phrase that best answers the question.

1. Why does magma tend to rise toward Earth's surface?
   A) It is more dense than surrounding rocks.
   B) It is more massive than surrounding rocks.
   C) It is cooler than surrounding rocks.
   D) It is less dense than surrounding rocks.

2. During metamorphism of granite into gneiss, what happens to minerals?
   A) They partly melt.
   B) They become new sediments.
   C) They grow smaller.
   D) They align into layers.

3. Which rock has large mineral grains?
   A) granite   C) obsidian
   B) basalt   D) pumice

4. What do igneous rocks form from?
   A) sediments   C) gravel
   B) mud   D) magma

5. What kind of rock is marble?
   A) foliated   C) intrusive
   B) nonfoliated   D) extrusive

6. What sedimentary rock is made of large, angular pieces of sediments?
   A) conglomerate   C) limestone
   B) breccia   D) chalk

7. Which of the following is an example of a detrital sedimentary rock?
   A) limestone   C) breccia
   B) evaporite   D) chalk

8. During what process are sediments pressed together?
   A) cooling   C) melting
   B) weathering   D) compaction

9. What is molten material at Earth's surface called?
   A) limestone   C) breccia
   B) lava   D) granite

10. Which of these is an organic sedimentary rock?
    A) coquina   C) rock salt
    B) sandstone   D) conglomerate

Thinking Critically

11. Granite, pumice, and scoria are igneous rocks. Why doesn't granite have airholes like the other two?

12. Why does marble rarely contain fossils?

13. Why is coquina classified as an organic rock with a granular texture?

14. Would you expect a quartzite or a sandstone to break more easily? Explain your answer.

15. Why are granitic igneous rocks lighter in color than basaltic rocks?

Developing Skills

16. **Comparing and Contrasting** Compare and contrast basaltic and granitic magmas.

17. **Forming Hypotheses**
    A geologist found a sequence of rocks in which 200-million-year-old shales were on top of 100-million-year-old sandstones. *Hypothesize how this could happen.*

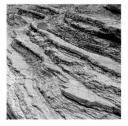

18. **Recognizing Cause and Effect** Explain the effects of pressure and temperature on shale.

19. Measuring in SI Assume that the conglomerate shown on the first page of the 2-page Activity is one half of its actual size. Determine the average length of the gravel in the rock.

20. Concept Mapping Copy and complete the concept map shown below. Use the following terms and phrases: *magma, sediments, igneous rock, sedimentary rock, metamorphic rock.* Add and label any missing arrows.

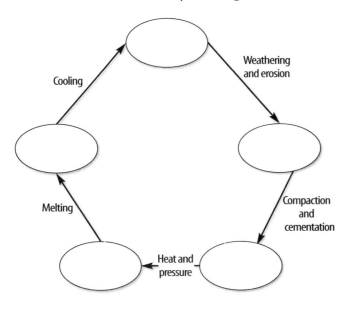

Performance Assessment

21. Poster Collect a group of small rocks. Make a poster that shows the classifications of rocks, and glue your rocks to the poster under the proper headings. Display your poster to the class. Describe your rocks and explain where you found them.

TECHNOLOGY

Go to the Glencoe Science Web site at **science.glencoe.com** or use the **Glencoe Science CD-ROM** for additional chapter assessment.

THE PRINCETON REVIEW **Test Practice**

Ellen did research on some common rocks. The information she gathered is listed in the table below.

Rock Properties and Uses		
Rocks	**Properties**	**Uses**
Granite	Hard; resistant to attack by air and natural waters	Building stones and monuments
Pumice	Lightweight and porous	Polishing mixtures and cleansers
Limestone	Soft; weathers in presence of natural waters	Gravel, concrete
Slate	Nonporous and splits into thin sheets	Roofing, paving, chalkboards

Study the table and answer the following questions.

1. According to the table, which rock would be a good choice to use as a grave marker in a cemetery?
A) granite
B) pumice
C) sandstone
D) slate

2. According to the information given, which rock would most likely be used in hand soap?
F) granite
G) pumice
H) sandstone
J) slate

3

Earth's Energy and Mineral Resources

Oil fields in Texas, like the one shown here, harbor some of the world's richest reserves of energy resources. In this chapter, you'll learn about many different types of resources that humans extract from Earth. You also will appreciate the importance of conserving these resources—especially those that are extracted faster than nature can produce them.

What do you think?

Science Journal Look at the picture below with a classmate. Discuss what this might be. Here's a hint: *It's on the cutting edge of oil exploration.* Write your answer in your Science Journal.

EXPLORE ACTIVITY

The physical properties of Earth materials determine how easily liquids and gases move through them. Geologists use these properties, in part, to predict where reserves of energy resources like petroleum or natural gas can be found.

Compare rock permeability

1. Obtain a sample of sandstone and a sample of shale from your teacher.
2. Make sure that your samples can be placed on a tabletop so that the sides facing up are reasonably flat and horizontal.
3. Place the two samples side by side in a shallow baking pan.
4. Using a dropper, place three drops of cooking oil on each sample.
5. For ten minutes, observe what happens to the oil on the samples.

Observe

Write your observations in your Science Journal. Infer which rock type might be a good reservoir for petroleum.

Before You Read

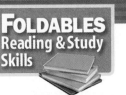

FOLDABLES
Reading & Study Skills

Making a Main Ideas Study Fold Make the following Foldable to help you identify the major topics about energy and mineral resources.

1. Place a sheet of paper in front of you so the long side is at the top. Fold the paper in half from the left side to the right side and then unfold.
2. Fold each side in to the centerfold line to divide the paper into fourths. Fold the paper in half from top to bottom and unfold.
3. Through the top thickness of paper, cut along both of the middle fold lines to form four tabs. Label the tabs *Nonrenewable Energy Resources, Inexhaustible Energy Resources, Renewable Energy Resources,* and *Mineral Resources.*
4. As you read the chapter, list examples on the front of the tabs and write about each type of resource under the tabs.

Nonrenewable Energy Resources

What **You'll Learn**

- **Identify** examples of nonrenewable energy resources.
- **Describe** the advantages and disadvantages of using fossil fuels.
- **Explain** the advantages and disadvantages of using nuclear energy.

Vocabulary

fossil fuel   natural gas
coal   reserve
oil   nuclear energy

Why **It's Important**

Nonrenewable resources should be conserved to ensure their presence for future generations.

Energy

The world's population relies on energy of all kinds. Energy is the ability to cause change. Some energy resources on Earth are being used faster than natural Earth processes can replace them. These resources are referred to as nonrenewable energy resources. Most of the energy resources used to generate electricity are nonrenewable.

Fossil Fuels

Nonrenewable energy resources include fossil fuels. **Fossil fuels** are fuels such as coal, oil, and natural gas that form from the remains of plants and other organisms that were buried and altered over millions of years. Coal is a sedimentary rock formed from the compacted and transformed remains of ancient plant matter. Oil is a liquid hydrocarbon that often is referred to as petroleum. Hydrocarbons are compounds that contain hydrogen and carbon atoms. Other naturally occurring hydrocarbons occur in the gas or semisolid states. Fossil fuels are processed to make gasoline for cars, to heat homes, and for many other uses, as shown in **Table 1.**

Table 1 Uses of Fossil Fuels	
Coal	■ To generate electricity
Oil	■ To produce gasoline and other fuels ■ As lubricants ■ To make plastics, home shingles, and other products
Natural Gas	■ To heat buildings ■ As a source of sulfur

Figure 1
This coal layer is located in Castle Gate, Utah. *Using the map and legend below, can you determine what type of coal it is?*

Legend:
- Anthracite
- Bituminous
- Lignite

CANADA

UNITED STATES

MEXICO

Coal The most abundant fossil fuel in the world is coal, shown in **Figure 1.** If the consumption of coal continues at the current rate, it is estimated that the coal supply will last for about another 250 years.

Coal is a rock that contains at least 50 percent plant remains. Coal begins to form when plants die in a swampy area. The dead plants are covered by more plants, water, and sediment, preventing atmospheric oxygen from coming into contact with the plant matter. The lack of atmospheric oxygen prevents the plant matter from decaying rapidly. Bacterial growth within the plant material causes a gradual breakdown of molecules in the plant tissue, leaving carbon and some impurities behind. This is the material that eventually will become coal after millions of years. Bacteria also cause the release of methane gas, carbon dioxide, ammonia, and water as the original plant matter breaks down.

☑ **Reading Check** *What happens to begin the formation of coal in a swampy area?*

Synthetic Fuels Unlike gasoline, which is refined from petroleum, other fuels called synthetic fuels are extracted from solid organic material. Synthetic fuels can be created from coal—a sedimentary rock containing hydrocarbons. The hydrocarbons are extracted from coal to form liquid and gaseous synthetic fuels. Liquid fuels can be processed to produce gasoline for automobiles and fuel oil for home heating. Gaseous synthetic fuels can be used to generate electricity and to heat buildings.

Life Science
INTEGRATION

The coal found in the eastern and midwestern United States formed from plants that lived in great swamps about 300 million years ago during the Pennsylvanian Period of geologic time. Research the Pennsylvanian Period to find out what types of plants lived in these swamps. Describe the plants in your Science Journal.

Stages of Coal Formation As decaying plant material loses gas and moisture, the concentration of carbon increases. The first step in this process, shown in **Figure 2,** results in the formation of peat. Peat is a layer of organic sediment. When peat burns, it releases large amounts of smoke because it has a high concentration of water and impurities.

As peat is buried under more sediment, it changes into lignite, which is a soft, brown coal with much less moisture. Heat and pressure produced by burial force water out of peat and concentrate carbon in the lignite. Lignite releases more energy and less smoke than peat when it is burned.

As the layers are buried deeper, bituminous coal, or soft coal, forms. Bituminous coal is compact, black, and brittle. It provides lots of heat energy when burned. Bituminous coal contains various levels of sulfur, which can pollute the environment.

If enough heat and pressure are applied to buried layers of bituminous coal, anthracite coal forms. Anthracite coal contains the highest amount of carbon of all forms of coal. Therefore, anthracite coal is the cleanest burning of all coals.

Figure 2
Coal is formed in four basic stages.

A Plant material accumulates in swamps and eventually forms a layer of peat.

B Over time, heat and pressure cause the peat to change into lignite coal.

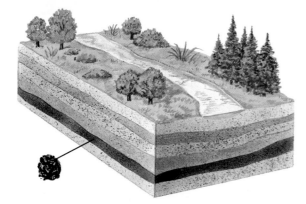

C As the lignite coal becomes buried by more sediments, heat and pressure change it into bituminous coal.

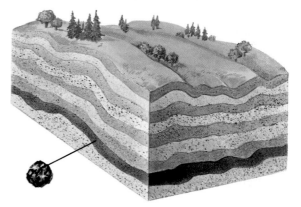

D When bituminous coal is heated and squeezed during metamorphism, anthracite coal forms.

Oil and Natural Gas

Coal isn't the only fossil fuel used to obtain energy. Two other fossil fuels that provide large quantities of the energy used today are oil and natural gas. **Oil** is a thick, black liquid formed from the buried remains of microscopic marine organisms. **Natural gas** forms under similar conditions and often with oil, but it forms in a gaseous state. Oil and natural gas are hydrocarbons. However, natural gas is composed of hydrocarbon molecules that are lighter than those in oil.

Residents of the United States burn vast quantities of oil and natural gas for daily energy requirements. As shown in **Figure 3,** Americans obtain most of their energy from these sources. Natural gas is used mostly for heating and cooking. Oil is used in many ways, including as heating oil, gasoline, lubricants, and in the manufacture of plastics and other important compounds.

Formation of Oil and Natural Gas

Most geologists agree that petroleum forms over millions of years from the remains of tiny marine organisms in ocean sediment. The process begins when marine organisms called plankton die and fall to the seafloor. Similar to the way that coal is buried, sediment is deposited over them. The temperature rises with depth in Earth, and increased heat eventually causes the dead plankton to change to oil and gas after they have been buried deeply by sediment.

Oil and natural gas often are found in layers of rock that have become tilted or folded. Because they are less dense than water, oil and natural gas are forced upward. Rock layers that are impermeable, such as shale, stop this upward movement. When this happens, a folded shale layer can trap the oil and natural gas below it. Such a trap for oil and gas is shown in **Figure 4.** The rock layer beneath the shale in which the petroleum and natural gas accumulate is called a reservoir rock.

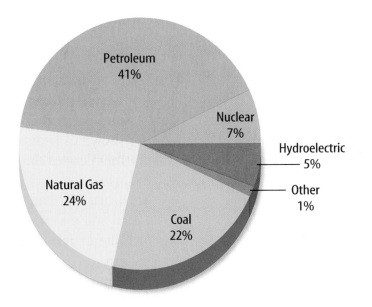

Figure 3
This graph shows the percentages of energy that the United States derives from various energy resources.

Figure 4
Oil and natural gas are fossil fuels formed by the burial of marine organisms. These fuels can be trapped and accumulate beneath Earth's surface.

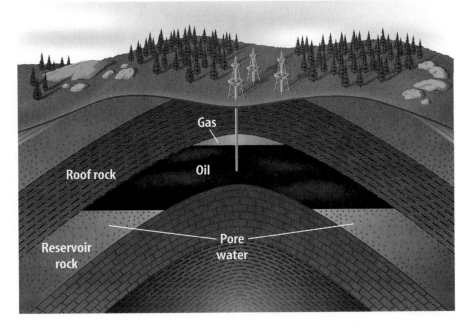

Removing Fossil Fuels from the Ground

Coal is removed from the ground using one of several methods of excavation. The two most common methods are strip mining, also called open-pit mining, and underground mining, shown in **Figure 5.** Oil and natural gas are removed by pumping them out of the ground.

Coal Mining During strip mining, shown in **Figure 5A,** layers of soil and rock above coal are removed and piled to one side. The exposed coal then is removed and loaded into trucks or trains and transported elsewhere. After the coal has been removed, mining companies often return the soil and rock to the open pit and cover it with topsoil. Trees and grass are planted in a process called land reclamation. If possible, animals native to the area are reintroduced. Strip mining is used only when the coal deposits are close to the surface.

In one method of underground coal mining, tunnels are dug and pillars of rock are left to support the rocks surrounding the tunnels. Two types of coal mines are drift mines and slope mines. Drift mining, shown in **Figure 5B,** is the removal of coal that is not close to Earth's surface through a horizontal opening in the side of a hill or mountain. In slope mining, an angled opening and air shaft are made in the side of a mountain to remove coal.

Figure 5
Coal is a fossil fuel that can be removed from Earth in many different ways.

A During strip mining, coal is accessed by removing the soil and rock above it.

B During underground coal mining, tunnels are made into the earth.
How do you think the coal is removed from these tunnels?

Drilling for Oil and Gas Oil and natural gas are fossil fuels that can be pumped from underground deposits. Geologists and engineers drill wells through rocks where these resources might be trapped, as shown in **Figure 6.** As the well is being drilled, it is lined with pipe to prevent it from caving in. When the drill bit reaches the rock layer containing oil, drilling is stopped. Equipment is installed to control the flow of oil. The surrounding rock then is fractured to allow oil and gas to flow into the well. The oil and gas are pumped to the surface.

> ✔ **Reading Check** *How are oil and natural gas brought to Earth's surface?*

Figure 6
Oil and natural gas are recovered from Earth by drilling deep wells.

Fossil Fuel Reserves

The amount of a fossil fuel that can be extracted at a profit using current technology is known as a **reserve.** This is not the same as a fossil fuel resource. A fossil fuel resource has fossil fuels that are concentrated enough that they can be extracted from Earth in useful amounts. However, a resource is not classified as a reserve unless the fuel can be extracted economically. What might cause a known fossil fuel resource to become classified as a reserve?

Methane Hydrates You have learned that current reserves of coal will last about 250 years. Enough natural gas is located in the United States to last about 60 more years. However, recent studies indicate that a new source of methane, which is the main component of natural gas, might be located beneath the seafloor. Icelike substances known as methane hydrates could provide tremendous reserves of methane.

Methane hydrates are stable molecules found hundreds of meters below sea level in ocean floor sediment. They form under conditions of relatively low temperatures and high pressures. The hydrocarbons are trapped within the cagelike structure of ice, as described in **Figure 7.** Scientists estimate that more carbon is contained in methane hydrates than in all current fossil fuel deposits combined. Large accumulations of methane hydrates are estimated to exist off the eastern coast of the United States. Can you imagine what it would mean to the world's energy supply if relatively clean-burning methane could be extracted economically from methane hydrates?

SCIENCE *Online*

Research Visit the Glencoe Science Web site at **science.glencoe.com** for more information about methane hydrates. Communicate to your class what you learn.

Figure 7

Reserves of fossil fuels—such as oil, coal, and natural gas—are limited and will one day be used up. Methane hydrates could be an alternative energy source. This icelike substance, background, has been discovered in ocean floor sediments and in permafrost regions worldwide. If scientists can harness this energy, the world's gas supply could be met for years to come.

Methane hydrates are highly flammable compounds made up of methane—the main component of natural gas—trapped in a cage of frozen water. Methane hydrates represent an enormous source of potential energy. However, they contain a greenhouse gas that might intensify global warming. More research is needed to determine how to safely extract them from the seafloor.

In the photo above, a Russian submersible explores a site in the North Atlantic that contains methane hydrate deposits.

Conserving Fossil Fuels Do you sometimes forget to turn off the lights when you walk out of a room? Wasteful habits might mean that electricity to run homes and industries will not always be as plentiful and cheap as it is today. Fossil fuels take millions of years to form and are used much faster than Earth processes can replenish them.

Today, coal provides about 25 percent of the energy that is used worldwide and 22 percent of the energy used in the United States. Oil and natural gas provide almost 61 percent of the world's energy and about 65 percent of the U.S. energy supply. At the rate these fuels are being used, they could run out some-day. How can this be avoided?

By remembering to turn off lights and appliances, you can avoid wasting fossil fuels. Another way to conserve fossil fuels is to make sure doors and windows are shut tightly during cold weather so heat doesn't leak out of your home. If you have air-conditioning, run it as little as possible. Ask the adults you live with if more insulation could be added to your home or if an insulated jacket could be put on the water heater.

Energy from Atoms

Most electricity in the United States is generated in power plants that use fossil fuels. However, alternate sources of energy exist. **Nuclear energy** is an alternate energy source produced from atomic reactions. When the nucleus of a heavy element is split, lighter elements form and energy is released. This energy can be used to light a home or power the submarines shown in **Figure 8.**

The splitting of heavy elements to produce energy is called nuclear fission. Nuclear fission is carried out in nuclear power plants using a type of uranium as fuel.

TRY AT HOME
Mini LAB

Practicing Energy Conservation

Procedure
1. Have an adult help you find the **electric meter** for your home and record the read-ing in your **Science Journal.**
2. Do this for several days, taking your meter readings at about the same time each day.
3. List things you and your family can do to reduce your electricity use.
4. Encourage your family to try some of the listed ideas for several days.

Analysis
1. Keep taking meter readings and infer whether the changes make any difference.
2. Have you and your family helped conserve energy?

A

Figure 8
Atoms can be a source of energy.
A These submarines are powered by nuclear fission. **B** Energy is given off when a larger atom, like uranium, splits into smaller atoms.

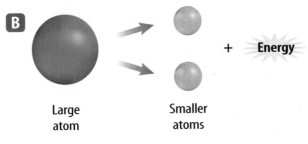

Large atom

Smaller atoms

+ Energy

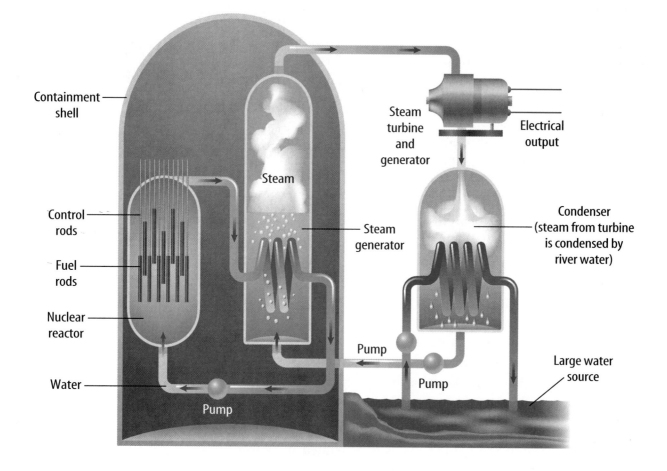

Containment shell

Control rods

Fuel rods

Nuclear reactor

Water

Pump

Steam

Steam turbine and generator

Electrical output

Steam generator

Condenser (steam from turbine is condensed by river water)

Pump

Pump

Large water source

Figure 9
Heat released in nuclear reactors produces steam, which in turn is used to produce electricity. This is an example of transforming nuclear energy into electrical energy.

Electricity from Nuclear Energy A nuclear power plant, shown in **Figure 9,** has a large chamber called a nuclear reactor. Within the nuclear reactor, uranium fuel rods sit in a pool of cooling water. Neutrons are fired into the fuel rods. When the uranium-235 atoms are hit, they break apart and fire out neutrons that hit other atoms, beginning a chain reaction. As each atom splits, it not only fires neutrons but also releases heat that is used to boil water to make steam. The steam drives a turbine, which turns a generator that produces electricity.

Reading Check *How is nuclear energy used to produce electricity?*

Nuclear energy from fission is considered to be a nonrenewable energy resource because it uses uranium-235 as fuel. A limited amount of uranium-235 is available for use. Another problem with nuclear energy is the waste material that it produces. Nuclear waste from power plants consists of highly radioactive elements formed by the fission process. Some of this waste will remain radioactive for thousands of years. The Environmental Protection Agency (EPA) has determined that nuclear waste must be stored safely and contained for at least 10,000 years before reentering the environment.

Fusion Environmental problems related to nuclear power could be eliminated if usable energy could be obtained from fusion. The Sun is a natural fusion power plant that provides energy for Earth and the solar system. Someday fusion also might provide energy for your home.

During fusion, materials of low mass are fused together to form a substance of higher mass. No fuel problem exists if the low-mass material is a commonly occurring substance. Also, if the end product is not radioactive, storing nuclear waste is not a problem. In fact, fusion of hydrogen into helium would satisfy both of these conditions. However, technologies do not currently exist to enable humans to fuse hydrogen into helium at reasonably low temperatures in a controlled manner. But research is being conducted, as shown in **Figure 10.** If this is accomplished, nuclear energy could be considered an inexhaustible fuel resource. You will learn the importance of inexhaustible and renewable energy resources in the next section.

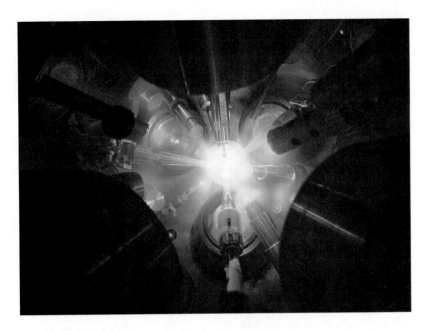

Figure 10
Lasers are used in research facilities to help people understand and control fusion.

Section 1 Assessment

1. Why are coal, oil, and natural gas considered to be fossil fuels?

2. Why are fossil fuels considered to be nonrenewable energy resources?

3. Describe similarities and differences among peat, lignite, bituminous coal, and anthracite coal.

4. How is nuclear energy obtained? What are two disadvantages of nuclear energy?

5. **Think Critically** Fossil fuels form in specific geologic environments. Why are you likely to find natural gas and oil deposits in the same location, but less likely to find coal and petroleum deposits at the same location?

Skill Builder Activities

6. **Making and Using Tables** Current energy consumption by source in the United States is as follows: *petroleum, 41 percent; natural gas, 24 percent; coal, 22 percent; nuclear energy, seven percent;* and *inexhaustible and renewable resources, six percent.* Make a bar graph of these data. **For more help, refer to the** Science Skill Handbook.

7. **Using Percentages** Using the data from question 6, what total percent of energy in the United States comes from petroleum and natural gas? How many times more energy is obtained from these sources than from coal? **For more help, refer to the** Math Skill Handbook.

Inexhaustible and Renewable Energy Resources

As You Read

What **You'll Learn**

- **Compare and contrast** inexhaustible and renewable energy resources.
- **Explain** why inexhaustible and renewable resources are used less than nonrenewable resources.

Vocabulary

solar energy   hydroelectric energy
wind farm   geothermal energy
          biomass energy

Why **It's Important**

As fossil fuel reserves continue to diminish, alternate energy resources will be needed.

Inexhaustible Energy Resources

How soon the world runs out of fossil fuels depends on how they are used and conserved. Fortunately, there are inexhaustible energy resources. These sources of energy are constant and will not run out in the future as fossil fuels will. Inexhaustible energy resources include the Sun, wind, water, and geothermal energy.

Energy from the Sun When you sit in the Sun, walk into the wind, or sail against an ocean current, you are experiencing the power of solar energy. **Solar energy** is energy from the Sun. You already know that the Sun's energy heats Earth, and it causes circulation in Earth's atmosphere and oceans. Ocean currents and global winds are examples of nature's use of solar energy. Thus, solar energy is used indirectly when the wind and some types of moving water are used to do work.

People can use solar energy in a passive way or in an active way. South-facing windows on buildings act as passive solar collectors, warming exposed rooms. Solar cells actively collect energy from the Sun and transform it into electricity. Solar cells were invented to generate electricity for satellites. Now they also are used to power calculators, streetlights, and experimental cars. Some people have installed solar energy cells on their roofs, as shown in **Figure 11**.

Figure 11
Solar panels, such as on this home in Laguna Niguel, California, can be used to collect inexhaustible solar energy to power appliances and heat water.

Figure 12
Wind farms are used to produce electricity.

Disadvantages of Solar Energy Solar energy is clean and inexhaustible, but it does have some disadvantages. Solar cells work less efficiently on cloudy days and cannot work at all at night. Some systems use batteries to store solar energy for use at night or on cloudy days, but it is difficult to store large amounts of energy in batteries. Worn out batteries also must be discarded. This can pollute the environment if not done properly.

Energy from Wind What is better to do on a warm, windy day than fly a kite? A strong wind can lift a kite high in the sky and whip it around. The pull of the wind is so great that you wonder if it will whip the kite right out of your hands. Wind is a source of energy. It was and still is used to power sailing ships. Windmills have used wind energy to grind corn and pump water. Today, windmills can be used to generate electricity. When a large number of windmills are placed in one area for the purpose of generating electricity, the area is called a **wind farm,** as shown in **Figure 12.**

Wind energy has advantages and disadvantages. Wind is non-polluting and free. It does little harm to the environment and produces no waste. However, only a few regions of the world have winds strong enough to generate electricity. Also, wind isn't steady. Sometimes it blows too hard and at other times it is too weak or stops entirely. For an area to use wind energy consistently, the area must have a persistent wind that blows at an appropriate speed.

Physics
INTEGRATION

Wind is an inexhaustible energy resource that is used by windmills to produce energy. As the blades rotate, turbines are turned to produce electricity. Find out which areas utilize wind farms and report in your Science Journal how much electricity is produced and what it is used for.

✔ **Reading Check** *Why are some regions better suited for wind farms than others?*

Energy from Water For a long time, waterwheels steadily spun next to streams and rivers. The energy in the flowing water powered the wheels that ground grain or cut lumber. More than a pretty picture, using a waterwheel in this way is an example of microhydropower. Microhydropower has been used throughout the world to do work.

Today energy from running water is used to generate electricity. Electricity produced by waterpower is called **hydroelectric energy.** To generate electricity from water running in a river, a large concrete dam is built to retain water, as illustrated in **Figure 13.** A lake forms behind the dam. As water is released, its force turns turbines at the base of the dam. The turbines then turn generators that make electricity.

At first it might appear that hydroelectric energy doesn't create any environmental problems and that the water is used with little additional cost. However, when dams are built, upstream lakes fill with sediment and downstream erosion increases. Land above the dam is flooded, and wildlife habitats are damaged.

Energy from Earth Erupting volcanoes and geysers like Old Faithful are examples of geothermal energy in action. The energy that causes volcanoes to erupt or water to shoot up as a geyser also can be used to generate electricity. Energy obtained by using hot magma or hot, dry rocks inside Earth is called **geothermal energy.**

Bodies of magma can heat large reservoirs of groundwater. Geothermal power plants use steam from the reservoirs to produce electricity, as shown in **Figure 14.** In a developing method, water becomes steam when it is pumped through broken, hot, dry rocks. The steam then is used to turn turbines that run generators to make electricity. The advantage of using hot, dry rocks is that they are found just about everywhere. Geothermal energy presently is being used in Hawaii and in parts of the western United States.

Figure 13
Hydroelectric power is important in many regions of the United States. **A** Hoover Dam was built on the Colorado River to supply electricity for a large area.
B The power of running water is converted to usable energy in a hydroelectric power plant.

A

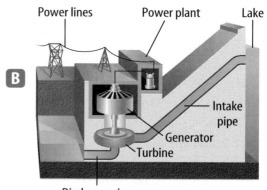

B

Power lines
Power plant
Lake
Intake pipe
Generator
Turbine
Discharge pipe

Figure 14
Geothermal energy is used to supply
electricity to industries and homes.

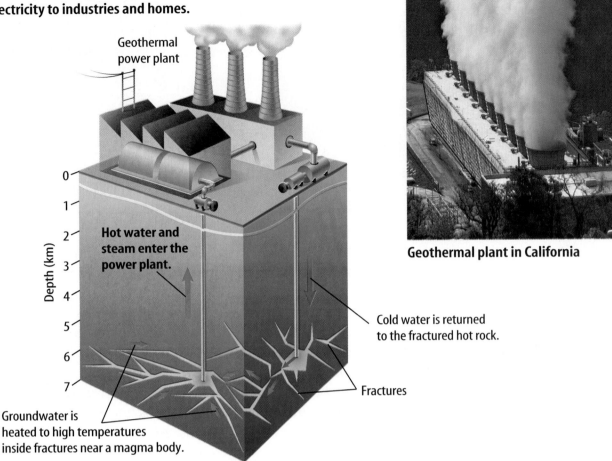

Geothermal
power plant

Depth (km)

0
1
2
3
4
5
6
7

**Hot water and
steam enter the
power plant.**

Cold water is returned
to the fractured hot rock.

Fractures

Groundwater is
heated to high temperatures
inside fractures near a magma body.

Geothermal plant in California

Renewable Energy Resources

Energy resources that can be replaced in nature or by
humans within a relatively short period of time are referred to
as renewable energy resources. This short period of time is
defined generally as being within a human life span. For example, trees can be considered a renewable energy resource. As one
tree is cut down, another can be planted in its place. The new
tree might be left untouched or harvested.

You have learned that most energy used in the United
States—about 90 percent—comes from fossil fuels, which are
nonrenewable energy resources. Next, you'll look at some
renewable energy resources and how they might fit into the
world's total energy needs now and in the future.

Biomass Energy

A major renewable energy resource is biomass materials.
Biomass energy is energy derived from burning organic material such as wood, alcohol, and garbage. The term *biomass* is
derived from the words *biological* and *mass*.

SCIENCE
Online

Research Visit the
Glencoe Science Web site at
science.glencoe.com for
more information about biomass energy. Communicate
to your class what you learn.

Figure 15
These campers are using wood, a renewable energy resource, to produce heat and light.

Energy from Wood If you've ever sat around a campfire, like the campers shown in **Figure 15,** or close to a wood-burning fireplace to keep warm, you have used energy from wood. The burning wood is releasing stored solar energy as heat energy. Humans have long used wood as an energy resource. Much of the world still cooks with wood. In fact, firewood is used more widely today than any other type of biomass fuel.

Using wood as a biomass fuel has its problems. Gases and small particles are released when wood is burned. These materials can pollute the air. When trees are cut down for firewood, natural habitats are destroyed. However, if proper conservation methods are employed or if tree farms are maintained specifically for use as fuel, energy from wood can be a part of future energy resources.

Energy from Alcohol Biomass fuel can be burned directly, such as when wood or peat is used for heating and cooking. However, it also can be transformed into other materials that might provide cleaner, more efficient fuels.

For example, during distillation, biomass fuel, such as corn, is changed to an alcohol such as ethanol. Ethanol then can be mixed with another fuel. When the other fuel is gasoline, the mixture is called gasohol. Gasohol can be used in the same way as gasoline, as shown in **Figure 16,** but it cuts down on the amount of fossil fuel needed to produce gasoline. Fluid biomass fuels are more efficient and have more uses than solid biomass fuels do.

The problem with this process is that presently, growing the corn and distilling the ethanol often uses more energy from burning fossil fuels than the amount of energy that is derived from burning ethanol. At present, biomass fuel is best used locally.

Figure 16
Gasohol sometimes is used to reduce dependence on fossil fuels.

✔ **Reading Check** *What are the drawbacks of biomass fuels?*

Energy from Garbage Every day humans throw away a tremendous amount of burnable garbage. As much as two thirds of what is thrown away could be burned. If more garbage were used for fuel, as shown in **Figure 17,** human dependence on fossil fuels would decrease. Burning garbage is a cheap source of energy and also helps reduce the amount of material that must be dumped into landfills.

Compared to other nations, the United States lags in the use of municipal waste as a renewable energy resource. For example, in some countries in Western Europe, as much as half of the waste generated is used for biomass fuel. When the garbage is burned, heat is produced, which turns water to steam. The steam turns turbines that run generators to produce electricity.

Unfortunately, some problems can be associated with using energy from garbage. Burning municipal waste can produce toxic ash residue and air pollution. Substances such as heavy metals could find their way into the smoke from garbage and thus into the atmosphere.

Figure 17
Garbage can be burned to produce electricity at trash-burning power plants such as this one in Virginia.

Section 2 Assessment

1. What are some advantages and disadvantages of using solar energy, wind energy, and hydroelectric energy?

2. What is the difference between inexhaustible and renewable energy resources? Give specific examples of each.

3. How is geothermal energy used to create electricity?

4. Why are nonrenewable resources used more than inexhaustible and renewable resources?

5. **Think Critically** How could nuclear energy, which normally is classified as a nonrenewable energy resource, be reclassified as an inexhaustible energy resource?

Skill Builder Activities

6. **Classifying** Classify the following energy resources as inexhaustible or renewable: *wood, solar energy, hydroelectric energy, geothermal energy, ethanol, garbage,* and *wind energy.* **For more help, refer to the** Science Skill Handbook.

7. **Using an Electronic Spreadsheet** Using the spreadsheet capabilities of a computer, make a table of energy resources. Include an example of how each resource could be used. Then write a description in your Science Journal about how you could reduce the use of energy resources at home. **For more help, refer to the** Technology Skill Handbook.

Activity

Soaking up Solar Energy

Winter clothing tends to be darker in color than summer clothing. The color of the material used in the clothing affects its ability to absorb energy. In this activity, you will use different colors of soil to study this effect.

Time and Temperature			
Time (min)	Temperature Dish A (°C)	Temperature Dish B (°C)	Temperature Dish C (°C)
0.0			
0.5			
1.0			
1.5			

What You'll Investigate
How does color affect the absorption of energy?

Materials
dry, black soil
dry, brown soil
dry, sandy, white soil
thermometers (3)
ring stand
graph paper
colored pencils (3)
metric ruler

clear glass or plastic dishes (3)
200-watt gooseneck lamp
*200-watt lamp with reflector and clamp
watch or clock with second hand
*stopwatch
*Alternate materials

Safety Precautions

WARNING: *Handle glass with care so as not to break it. Wear thermal mitts when handling the light source.*

Goals
■ **Determine** whether color has an effect on the absorption of solar energy.
■ **Relate** the concept of whether color affects absorption to other applications.

Procedure
1. Fill each plastic dish with a different color of soil to a depth of 2.5 cm.

2. Arrange the dishes close together on your desk and place a thermometer in each dish. Be sure to cover the thermometer bulb in each dish completely with the soil.

3. Position the lamp over all three dishes.

4. **Design** a data table for your observations similar to the sample table above. You will need to read the temperature of each dish every 30 s for 20 min after the light is turned on.

5. Turn on the light and begin your experiment.

6. Use the data to construct a graph. Time should be plotted on the horizontal axis and temperature on the vertical axis. Use a different colored pencil to plot the data for each type of soil, or use a computer to design a graph that illustrates your data.

Conclude and Apply
1. Which soil had the greatest temperature change? The least?

2. Why do the curves on the graph flatten?

3. Why do flat-plate solar collectors have black plates behind the water pipes?

4. How does the color of a material affect its ability to absorb energy?

5. Why is most winter clothing darker in color than summer clothing?

Mineral Resources

Metallic Mineral Resources

If your room at home is anything like the one shown in **Figure 18,** you will find many metal items. Metals are obtained from Earth materials called metallic mineral resources. A **mineral resource** is a deposit of useful minerals. See how many metals you can find. Is there anything in your room that contains iron? What about the metal in the frame of your bed? Is it made of iron? If so, the iron might have come from the mineral hematite. What about the framing around the windows in your room? Is it aluminum? Aluminum, like that in a soft-drink can, comes from a mixture of minerals known as bauxite. Many minerals contain these and other useful elements. Which minerals are mined as sources for the materials you use every day?

Ores Deposits in which a mineral or minerals exist in large enough amounts to be mined at a profit are called **ores.** Generally, the term *ore* is used for metallic deposits, but this is not always the case. The hematite that was mentioned earlier as an iron ore and the bauxite that was mentioned earlier as an aluminum ore are metallic ores.

✔ Reading Check *What is an ore?*

As You Read

What You'll Learn

- **Explain** the conditions needed for a mineral to be classified as an ore.
- **Describe** how market conditions can cause a mineral to lose its value as an ore.
- **Compare and contrast** metallic and nonmetallic mineral resources.

Vocabulary
mineral resource
ore
recycling

Why It's Important
Many products you use are made from mineral resources.

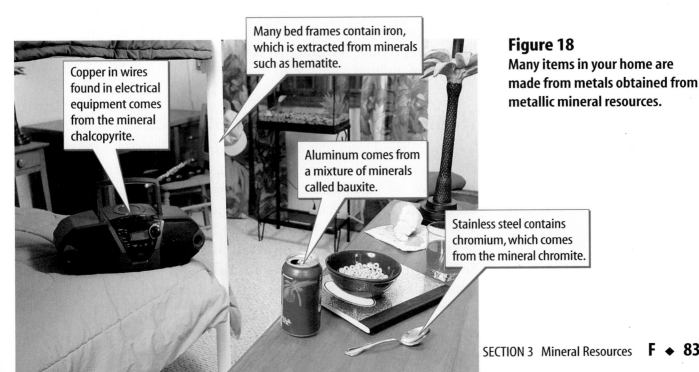

Many bed frames contain iron, which is extracted from minerals such as hematite.

Copper in wires found in electrical equipment comes from the mineral chalcopyrite.

Aluminum comes from a mixture of minerals called bauxite.

Stainless steel contains chromium, which comes from the mineral chromite.

Figure 18
Many items in your home are made from metals obtained from metallic mineral resources.

Figure 19
Iron ores are smelted to produce nearly pure iron. *What could this iron be used for?*

Economic Effects When is a mineral deposit considered an ore? The mineral in question must be in demand. Enough of it must be present in the deposit to make it worth removing. Some mining operations are profitable only if a large amount of the mineral is needed. It also must be fairly easy to separate the mineral from the material in which it is found. If any one of these conditions isn't met, the deposit might not be considered an ore.

Supply and demand is an important part of life. You might have noticed that when the supply of fresh fruit is down, the price you pay for it at the store goes up. Economic factors largely determine what an ore is.

Refining Ore The process of extracting a useful substance from an ore involves two operations—concentrating and refining. After a metallic ore is mined from Earth's crust, it is crushed and the waste rock is removed. The waste rock that must be removed before a mineral can be used is called gangue (GANG).

Refining produces a pure or nearly pure substance from ore. For example, iron can be concentrated from the ore hematite, which is composed of iron oxide. The concentrated ore then is refined to be as close to pure iron as possible. One method of refining is smelting, illustrated in **Figure 19**. Smelting is a chemical process that removes unwanted elements from the metal that is being processed. During one smelting process, a concentrated ore of iron is heated with a specific chemical. The chemical combines with oxygen in the iron oxide, resulting in pure iron. Note that one resource, fossil fuel, is burned to produce the heat that is needed to obtain the finished product of another resource, in this case iron.

Nonmetallic Mineral Resources

Any mineral resources not used as fuels or as sources of metals are nonmetallic mineral resources. These resources are mined for the nonmetallic elements contained in them and for the specific physical and chemical properties they have. Generally, nonmetallic mineral resources can be divided into two different groups—industrial minerals and building materials. Some materials, such as limestone, belong to both groups of nonmetallic mineral resources, and others are specific to one group or the other.

Industrial Minerals Many useful chemicals are obtained from industrial minerals. Sandstone is a source of silica (SiO_2), which is a compound that is used to make glass. Some industrial minerals are processed to make fertilizers for farms and gardens. For example, sylvite, a mineral that forms when seawater evaporates, is used to make potassium fertilizer.

Many people enjoy a little sprinkle of salt on french fries and pretzels. Table salt is a product derived from halite, a nonmetallic mineral resource. Halite also is used to help melt ice on roads and sidewalks during winter and to help soften water.

Other industrial minerals are useful because of their characteristic physical properties. For example, abrasives are made from deposits of corundum and garnet. Both of these minerals are hard and able to scratch most other materials they come into contact with. Small particles of garnet can be glued onto a sheet of heavy paper to make abrasive sandpaper. **Figure 20** illustrates just a few ways in which nonmetallic mineral resources help make your life more convenient.

Figure 20
You benefit from the use of industrial minerals every day.

A Road salt melts ice on streets.

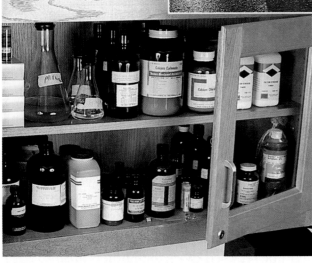
B Many important chemicals are made from industrial minerals.

C An industrial mineral called trona is important for making glass.

What stones are used in buildings where you live? To find out more about building stones, see the **Building Stones Field Guide** at the back of the book.

Building Materials One of the most important nonmetallic mineral resources is aggregate. Aggregate is composed of crushed stone or a mixture of gravel and sand and has many uses in the building industry. For example, aggregates can be mixed with cement and water to form concrete. Quality concrete is vital to the building industry. Limestone also has industrial uses. It is used as paving stone and as part of concrete mixtures. Have you ever seen the crushed rock in a walking path or driveway? The individual pieces might be crushed limestone. Gypsum, a mineral that forms when seawater evaporates, is soft and lightweight and is used in the production of plaster and wallboard. If you handle a piece of broken plaster or wallboard, note its appearance, which is similar to the mineral gypsum.

Rock also is used as building stone. You might know of buildings in your region that are made from granite, limestone, or sandstone. These rocks and others are quarried and cut into blocks and sheets. The pieces then can be used to construct buildings. Some rock also is used to sculpt statues and other pieces of art.

✓ **Reading Check** *What are some important nonmetallic mineral resources?*

Problem-Solving Activity

Why should you recycle?

Recycling in the United States has become a way of life. In 2000, 88 percent of Americans participated in recycling. Recycling is important because it saves precious raw materials and energy. Recycling aluminum saves 95 percent of the energy required to obtain it from its ore. Recycling steel saves up to 74 percent in energy costs, and recycling glass saves up to 22 percent.

Recycling Rates in the United States			
Material	**1995 (%)**	**1997 (%)**	**2000 (%) (estimated)**
Glass	24.5	24.3	29–33
Steel	36.5	38.4	41–46
Aluminum	34.6	31.2	37–39
Plastics	5.3	5.2	6–7

Identifying the Problem

The following table includes materials that currently are being recycled and rates of recycling for the years 1995, 1997, and 2000. Examine the table to determine materials for which recycling increased or decreased between 1995 and 2000.

Solving the problem

1. Has the recycling of materials increased or decreased over time? Which materials are recycled most? Which materials are recycled least? Suggest reasons why some materials might be recycled more than others are.
2. How can recycling benefit society? Explain your answer.

Recycling Mineral Resources

Mineral resources are nonrenewable. You've learned that nonrenewable resources are those that Earth processes cannot replace within an average human's lifetime. Most mineral resources take millions of years to form. Have you ever thrown away an empty soft-drink can? Many people do. These cans become solid waste. Wouldn't it be better if these cans and other items made from mineral resources were recycled into new items?

Recycling is using old materials to make new ones. Recycling has many advantages. It reduces the demand for new mineral resources. The recycling process often uses less energy than it takes to obtain new material. Because supplies of some minerals might become limited in the future, recycling could be required to meet needs for certain materials, as shown in **Figure 21.**

Recycling also can be a profitable experience. Some companies purchase scrap metal and empty soft-drink cans for the aluminum and tin content. The seller receives a small amount of money for turning in the material. Schools and other groups earn money by recycling soft-drink cans.

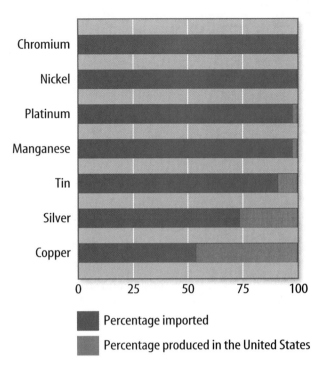

■ Percentage imported
■ Percentage produced in the United States

Figure 21
The United States produces only a small percentage of the metallic resources it consumes.

Section 3 Assessment

1. How are metals obtained from metallic mineral resources used in your home and school? Which of these products could be recycled easily?

2. What characteristics define a mineral deposit as an ore?

3. List two industrial uses for nonmetallic mineral resources.

4. How can supply and demand of a material cause a mineral to become an ore?

5. **Think Critically** Gangue is waste rock remaining after a mineral ore is removed. Why is gangue sometimes reprocessed?

Skill Builder Activities

6. **Classifying** Classify the following mineral resources as metallic or nonmetallic: *hematite, limestone, bauxite, sandstone, garnet,* and *chalcopyrite.* Explain why you classified each one as you did. **For more help, refer to the** Science Skill Handbook.

7. **Communicating** Research the element titanium and its uses. Write a description in your Science Journal explaining the importance of this element and the mineral resources it comes from. Explain why this element is used in the ways that you found. **For more help, refer to the** Science Skill Handbook.

Home Sweet Home

As fossil fuel supplies continue to be depleted, an increasing U.S. population has recognized the need for alternative energy sources. United States residents might be forced to consider using renewable and inexhaustible energy resources to meet some of their energy needs. The need for energy-efficient housing is more relevant now than ever before. A designer of energy-efficient homes considers proper design and structure, a well chosen building site with wise material selection, and selection of efficient energy generation systems to power the home. Energy-efficient housing uses less energy and produces fewer pollutants.

Recognize the Problem

What does the floor plan, building plan, or a model of an energy efficient home look like?

Thinking Critically

How and where should your house be designed and built to use the alternative energy resources you've chosen efficiently?

Goals
- ■ **Research** various renewable and inexhaustible energy resources available to use in the home.
- ■ **Design** blueprints for an energy-efficient home and/or design and build a model of an energy-efficient home.

Possible Materials
paper
ruler
pencils
cardboard
glue
aluminum foil

Data Source
SCIENCE *Online* Go to the Glencoe Science Web site at **science.glencoe.com** for more information about designing an energy-efficient home.

FOR SALE
AWARD WINNING
PASSIVE SOLAR HOUSE

Planning the Model

Plan

1. **Research** current information about energy-efficient homes.

2. **Research** inexhaustible energy resources such as wind, hydroelectric power, or solar power, as well as energy conservation. Decide which energy resources are most efficient for your home design.

3. Decide where your house should be built to use energy efficiently.

4. Decide how your house will be laid out and draw mock blueprints for your home. Highlight energy issues such as where solar panels can be placed.

5. Build a model of your energy-efficient home.

Do

1. Ask your peers for input on your home. As you research, become an expert in one area of alternative energy generation and share your information with your classmates.

2. **Compare** your home's design to energy-efficient homes you learn about through your research.

Making the Model

1. Think about how most of the energy in a home is used. Remember as you plan your home that energy-efficient homes not only generate energy—they also use it more efficiently.

2. Carefully consider where your home should be built. For instance, if you plan to use wind power, will your house be built in an area that receives adequate wind?

3. Be sure to plan for backup energy generation. For instance, if you plan to use mostly solar energy, what will you do if it's a cloudy day?

Analyzing and Applying Results

1. Devise a budget for building your home. Could your energy-efficient home be built at a reasonable price? Could anyone afford to build it?

2. Create a list of pro and con statements about the use of energy-efficient homes. Why aren't inexhaustible and renewable energy sources widely used in homes today?

*C*ommunicating
Your Data

Present your model to the class. **Explain** which energy resources you chose to use in your home and why. Have an open house. Take prospective home owners/classmates on a tour of your home and sell it.

Oil's Well That

What if you went out to your backyard, started digging a hole, and all of a sudden oil spurted out of the ground? Dollar signs might flash before your eyes.

It wasn't quite that exciting for Charles Tripp. Tripp, a Canadian, is credited with being the first person to strike oil. And he wasn't even looking for what has become known as "black gold."

In 1851, Tripp built a factory in Ontario, Canada, not far from Lake Erie. He used a natural, black, thick, sticky substance that could be found nearby to make asphalt for paving roads and to construct buildings. He also sold the product to waterproof boats.

The Titusville, Pennsylvania, oil well drilled by Edwin Drake. This photo was taken in 1864.

In 1855, Tripp dug a well looking for fresh-water for his factory. After digging just 2 m or so, he unexpectedly came upon liquid. It wasn't clear, clean, and delicious; it was smelly, thick, and black. You guessed it—oil! Tripp didn't understand the importance of his find and how it might help his business. Nor did he realize that it might make him rich someday. So, two years after his accidental discovery, Tripp sold his company to James Williams.

In 1858, Williams continued to search for water for the factory, but, as luck would have it, diggers kept finding oil. So workers barreled up the oil and sold it as liquid grease for the train industry. By 1861, 1.5 million L of crude oil had been barreled and shipped. The finds also led to a short boom in oil production in the area.

Some people argue that the first oil well in North America was in Titusville, Pennsylvania, when Edwin Drake hit oil in 1859. However, most historians agree that Williams was first in 1858. But they also agree that it was Edwin Drake's discovery that led to the growth of the oil industry. So, Drake and Williams can share the credit!

Ends

Some people used TNT to search for oil. This photo was taken in 1943.

The accidental discovery of a gloppy black substance helped change the planet

Well

Today, many oil companies are drilling beneath the sea for oil.

CONNECTIONS Make a Graph Research the leading oil-producing nations and make a bar graph of the top five producers. Go further and research how rising and falling prices of crude oil can affect the U.S. and world economies. Share your findings with your class.

SCIENCE *Online*

For more information, visit science.glencoe.com.

Reviewing Main Ideas

Section 1 Nonrenewable Energy Resources

1. Fossil fuels are considered to be non-renewable energy resources.

2. The higher the concentration of carbon in coal is, the cleaner it burns. *What might the coal in this truck be used for?*

3. Oil and natural gas form from altered and buried marine organisms and often are found near one another.

4. Nuclear energy is obtained from the fission of heavy isotopes.

Section 2 Inexhaustible and Renewable Energy Resources

1. Inexhaustible energy resources—solar energy, wind energy, water energy, and geothermal energy—are constant and will not run out. *What are some advantages of using solar panels to produce electricity?*

2. Renewable energy resources are replaced within a relatively short period of time.

3. Biomass energy is derived from organic material such as wood and corn.

Section 3 Mineral Resources

1. Metallic mineral resources provide metals when they are processed.

2. Ores are mineral resources that contain a usable substance that can be mined at a profit.

3. Smelting is a chemical process that removes unwanted elements from a metal that is being processed.

4. Nonmetallic mineral resources are classified as industrial minerals or building materials. *Is the aggregate that is being produced here an industrial mineral or a building material?*

FOLDABLES
Reading & Study Skills

After You Read

Circle the resources you think are the most valuable on your Foldable. Explain why and write about similarities and differences among the resources.

Visualizing Main Ideas

Fill in the following table that lists advantages and disadvantages of different energy resources.

Energy Resources		
Resource	**Advantages**	**Disadvantages**
Fossil Fuels		
Nuclear Energy		
Solar Energy		
Wind Energy		
Geothermal Energy		
Biomass Fuel		

Vocabulary Review

Vocabulary Words

a. biomass energy
b. coal
c. fossil fuel
d. geothermal energy
e. hydroelectric energy
f. mineral resource
g. natural gas
h. nuclear energy
i. oil
j. ore
k. recycling
l. reserve
m. solar energy
n. wind farm

THE PRINCETON REVIEW **Study Tip**

To understand the information that a graph is trying to communicate, write a sentence that describes the relationship between the axes.

Using Vocabulary

Each phrase below describes a vocabulary word from the list. Write the word that matches the phrase describing it.

1. mineral resource that can be mined at a profit

2. fuel that is composed mainly of the remains of dead plants

3. method of conservation in which items are processed to be used again

4. inexhaustible energy resource that is used to power the *Hubble Space Telescope*

5. energy resource that is based on the fission of atoms

6. liquid from remains of marine organisms

Chapter ③ Assessment

Checking Concepts

Choose the word or phrase that best answers the question.

1. Which of the following has the highest content of carbon?
   A) peat
   B) lignite
   C) bituminous coal
   D) anthracite coal

2. Which of the following is the first step in the evolution of coal?
   A) formation of peat
   B) formation of lignite
   C) formation of bituminous coal
   D) formation of anthracite coal

3. Which of the following is an example of a fossil fuel?
   A) wind
   B) water
   C) natural gas
   D) uranium-235

4. What is the waste material that must be separated from an ore?
   A) smelter
   B) mineral resource
   C) gangue
   D) petroleum

5. What common rock structure can trap oil and natural gas under it?
   A) folded rock
   B) sandstone rock
   C) porous rock
   D) unconsolidated rock

6. Which type of energy resource uses large dams in a river?
   A) wind
   B) nuclear
   C) hydroelectric
   D) solar

7. What is a region where many windmills are located in order to generate electricity from wind called?
   A) wind farm
   B) hydroelectric dam
   C) oil well
   D) steam-driven turbine

8. Which of the following is a deposit of hematite that can be mined at a profit?
   A) ore
   B) anthracite
   C) gangue
   D) energy resource

9. What is an important use of petroleum?
   A) making plaster
   B) making glass
   C) as abrasives
   D) making gasoline

10. Which of the following is a nonrenewable energy resource?
    A) water
    B) wind
    C) geothermal
    D) petroleum

Thinking Critically

11. Explain how solar energy becomes stored in plants and other organisms and is released later when fossil fuels are burned.

12. Describe the major problems associated with generating electricity using nuclear power plants.

13. Why is wind considered to be an inexhaustible energy resource?

14. Which type of energy resources are considered to be biomass fuels?

15. What conditions could occur to cause gangue to be reclassified as an ore?

Developing Skills

16. **Predicting** If a well were drilled into a rock layer containing petroleum, natural gas, and water, which substance would be encountered first? Explain.

17. **Comparing and Contrasting** Create a table comparing and contrasting solar energy and wind energy.

18. Concept Mapping Complete the following concept map about mineral resources.

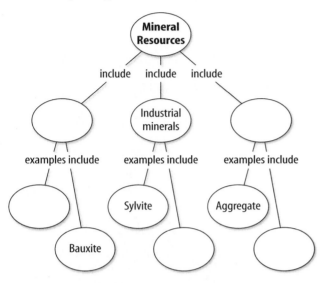

19. Making Models Make a model of a house that has been built to use passive solar energy.

Performance Assessment

20. Oral Presentation Research the latest information on inexhaustible energy resources. Develop a presentation that tries to persuade consumers to reduce their use of fossil fuels and to use inexhaustible energy resources instead.

21. Letter Write a letter to the Department of Energy asking how usable energy might be obtained from methane hydrates in the future. Also inquire about methods to extract methane hydrates.

TECHNOLOGY

Go to the Glencoe Science Web site at **science.glencoe.com** or use the **Glencoe Science CD-ROM** for additional chapter assessment.

THE PRINCETON REVIEW **Test Practice**

The graph below shows the average abundance of some important metals in Earth's crust.

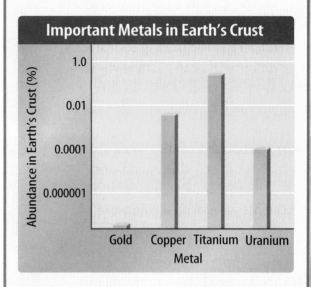

Study the bar graph and answer the following questions.

1. After examining this graph, you should conclude that these important metals _____.
   A) are concentrated in Earth's crust
   B) are valuable for jewelry
   C) are mined out
   D) are generally rare in Earth's crust

2. Which of the following best explains the occurrence of rich ores of these metals?
   F) The metals are concentrated in some places by Earth processes.
   G) Metal-rich meteorites frequently strike Earth.
   H) Some metals are more abundant in Earth's core.
   J) Percentages often are misleading.

Plate Tectonics

Characterized by volcanoes and scenic vistas, the East African Rift Valley marks a place where Earth's crust is being pulled apart. If the pulling continues over millions of years, Africa will separate into two landmasses. In this chapter, you'll learn about Rift Valleys and other features explained by the theory of plate tectonics. You'll also learn about the fossil, climate, and rock clues that indicate that Earth's continents have drifted over time.

What do you think?

Science Journal Look at the picture below with a classmate. Discuss what you think this might be or what is happening. Here's a hint: *A river runs through this dog leg.* Write your answer or best guess in your Science Journal.

Can you imagine a giant landmass that broke into many separate continents and Earth scientists working to reconstruct Earth's past? Do this activity to learn about clues that can be used to reassemble a supercontinent.

Reassemble an image

1. Collect interesting photographs from an old magazine.

2. You and a partner each select one photo, but don't show them to each other. Then each of you cut your photos into pieces no smaller than about 5 cm or 6 cm.

3. Trade your cut-up photo for your partner's.

4. Observe the pieces, and reassemble the photograph your partner has cut up.

Observe

In your Science Journal, describe the characteristics of the cut-up photograph that helped you put the image back together. Think of other examples in which characteristics of objects are used to match them up with other objects.

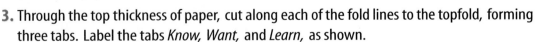

Before You Read

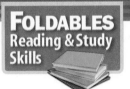

FOLDABLES
Reading & Study Skills

Making a Know-Want-Learn Study Fold It would be helpful to identify what you already know and what you want to know. Make the following Foldable to help you focus on reading about plate tectonics.

1. Place a sheet of paper in front of you so the long side is at the top. Fold the paper in half from top to bottom.

2. Fold both sides in to divide the paper into thirds. Unfold the paper so three sections show.

3. Through the top thickness of paper, cut along each of the fold lines to the topfold, forming three tabs. Label the tabs *Know, Want,* and *Learn,* as shown.

4. Before you read the chapter, write what you know about plate tectonics under the left tab and what you want to know under the middle tab.

5. As you read the chapter, write what you learn about plate tectonics under the right tab.

Continental Drift

As You Read

What **You'll Learn**

- **Describe** the hypothesis of continental drift.
- **Identify** evidence supporting continental drift.

Vocabulary
continental drift
Pangaea

Why **It's Important**
The hypothesis of continental drift led to plate tectonics—a theory that explains many processes in Earth.

Figure 1
This illustration represents how the continents once were joined to form Pangaea. This fitting together of continents according to shape is not the only evidence supporting the past existence of Pangaea.

Evidence for Continental Drift

If you look at a map of Earth's surface, you can see that the edges of some continents look as though they could fit together like a puzzle. Other people also have noticed this fact. For example, Dutch mapmaker Abraham Ortelius noted the fit between the coastlines of South America and Africa more than 400 years ago.

Pangaea German meteorologist Alfred Wegener (VEG nur) thought that the fit of the continents wasn't just a coincidence. He suggested that all the continents were joined together at some time in the past. In a 1912 lecture, he proposed the hypothesis of continental drift. According to the hypothesis of **continental drift,** continents have moved slowly to their current locations. Wegener suggested that all continents once were connected as one large landmass, shown in **Figure 1,** that broke apart about 200 million years ago. He called this large landmass **Pangaea** (pan JEE uh), which means "all land."

✔ **Reading Check** *Who proposed continental drift?*

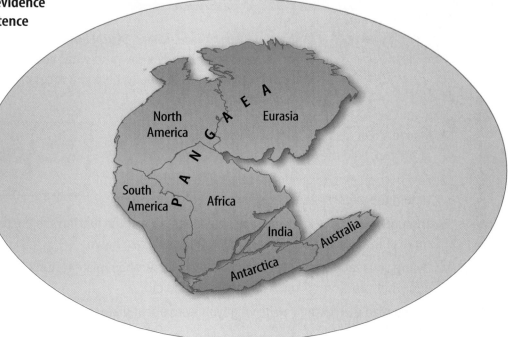

A Controversial Idea Wegener's ideas about continental drift were controversial. It wasn't until long after Wegener's death in 1930 that his basic hypothesis was accepted. The evidence Wegener presented hadn't been enough to convince many people during his lifetime. He was unable to explain exactly how the continents drifted apart. He proposed that the continents plowed through the ocean floor, driven by the spin of Earth. Physicists and geologists of the time strongly disagreed with Wegener's explanation. They pointed out that continental drift would not be necessary to explain many of Wegener's observations. Other important observations that came later eventually supported Wegener's earlier evidence.

Fossil Clues Besides the puzzlelike fit of the continents, fossils provided support for continental drift. Fossils of the reptile *Mesosaurus* have been found in South America and Africa, as shown in **Figure 2.** This swimming reptile lived in freshwater and on land. How could fossils of *Mesosaurus* be found on land areas separated by a large ocean of salt water? It probably couldn't swim between the continents. Wegener hypothesized that this reptile lived on both continents when they were joined.

✔ **Reading Check** *How do Mesosaurus fossils support the past existence of Pangaea?*

SCIENCE *Online*

Research Visit the Glencoe Science Web site at **science.glencoe.com** for more information about the continental drift hypothesis. Communicate to your class what you learn.

Figure 2
Fossil remains of plants and animals that lived in Pangaea have been found on more than one continent. *How do the locations of* Glossopteris, Mesosaurus, Kannemeyerid, Labyrinthodont, *and other fossils support Wegener's hypothesis of continental drift?*

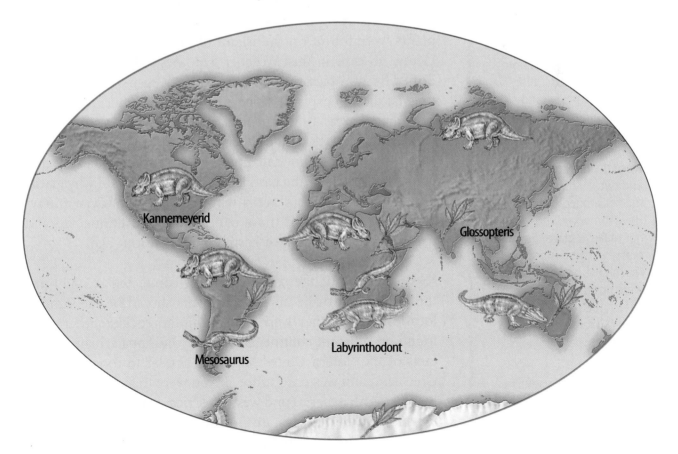

Figure 3
This fossil plant, *Glossopteris,* grew in a temperate climate.

Interpreting Fossil Data

Procedure

1. Build a three-layer landmass using **clay or modeling dough.**
2. Mold the clay into mountain ranges.
3. Place similar **"fossils"** into the clay at various locations around the landmass.
4. Form five continents from the one landmass. Also, form two smaller landmasses out of different clay with different mountain ranges and fossils.
5. Place the five continents and two smaller landmasses around the room.
6. Have someone who did not make or place the landmasses make a model that shows how they once were positioned.
7. Return the clay to its container so it can be used again.

Analysis

What clues were useful in reconstructing the original landmass?

A Widespread Plant Another fossil that supports the hypothesis of continental drift is *Glossopteris* (glahs AHP tur us). **Figure 3** shows this fossil plant, which has been found in Africa, Australia, India, South America, and Antarctica. The presence of *Glossopteris* in so many areas also supported Wegener's idea that all of these regions once were connected and had similar climates.

Climate Clues Wegener used continental drift to explain evidence of changing climates. For example, fossils of warm-weather plants were found on the island of Spitsbergen in the Arctic Ocean. To explain this, Wegener hypothesized that Spitsbergen drifted from tropical regions to the arctic. Wegener also used continental drift to explain evidence of glaciers found in temperate and tropical areas. Glacial deposits and rock surfaces scoured and polished by glaciers are found in South America, Africa, India, and Australia. This shows that parts of these continents were covered with glaciers in the past. How could you explain why glacial deposits are found in areas where no glaciers exist today? Wegener thought that these continents were connected and partly covered with ice near Earth's south pole long ago.

Rock Clues If the continents were connected at one time, then rocks that make up the continents should be the same in locations where they were joined. Similar rock structures are found on different continents. Parts of the Appalachian Mountains of the eastern United States are similar to those found in Greenland and western Europe. If you were to study rocks from eastern South America and western Africa, you would find other rock structures that also are similar. Rock clues like these support the idea that the continents were connected in the past.

250 million years ago

135 million years ago

Present day

How could continents drift?

Although Wegener provided evidence to support his hypothesis of continental drift, he couldn't explain how, when, or why these changes, shown in **Figure 4,** took place. The idea suggested that lower-density, continental material somehow had to plow through higher-density, ocean-floor material. The force behind this plowing was thought to be the spin of Earth on its axis—a notion that was quickly rejected by physicists. Because other scientists could not provide explanations either, Wegener's idea of continental drift was initially rejected. The idea was so radically different at that time that most people closed their minds to it.

Rock, fossil, and climate clues were the main types of evidence for continental drift. After Wegener's death, more clues were found, largely because of advances in technology, and new ideas that related to continental drift were developed. You'll learn about one of these new ideas, seafloor spreading, in the next section. Seafloor spreading helped provide an explanation of how the continents could move.

Figure 4
These computer models show the probable course the continents have taken. On the far left is their position 250 million years ago. In the middle is their position 135 million years ago. At right is their current position.

Section 1 Assessment

1. Why were Wegener's ideas about continental drift initially rejected?
2. How did Wegener use climate clues to support his hypothesis of continental drift?
3. What rock clues were used to support the hypothesis of continental drift?
4. In what ways do fossils help support the hypothesis of continental drift?
5. **Think Critically** Why would you expect to see similar rocks and rock structures on two landmasses that were connected at one time?

Skill Builder Activities

6. **Comparing and Contrasting** Compare and contrast the locations of fossils of the temperate plant *Glossopteris,* as shown in **Figure 2,** with the climate that exists at each location today. **For more help, refer to the** Science Skill Handbook.
7. **Communicating** Imagine that you are Alfred Wegener in the year 1912. In your Science Journal, write a letter to another scientist explaining your idea about continental drift. Try to convince this scientist that your hypothesis is correct. **For more help, refer to the** Science Skill Handbook.

Seafloor Spreading

As You Read

What You'll Learn

- **Explain** seafloor spreading.
- **Recognize** how age and magnetic clues support seafloor spreading.

Vocabulary
seafloor spreading

Why It's Important
Seafloor spreading helps explain how continents moved apart.

Figure 5
As the seafloor spreads apart at a mid-ocean ridge, new seafloor is created. The older seafloor moves away from the ridge in opposite directions.

Mapping the Ocean Floor

If you were to lower a rope from a boat until it reached the seafloor, you could record the depth of the ocean at that particular point. In how many different locations would you have to do this to create an accurate map of the seafloor? This is exactly how it was done until World War I, when the use of sound waves was introduced to detect submarines. During the 1940s and 1950s, scientists began using sound waves on moving ships to map large areas of the ocean floor in detail. Sound waves echo off the ocean bottom—the longer the sound waves take to return to the ship, the deeper the water is.

Using sound waves, researchers discovered an underwater system of ridges, or mountains, and valleys like those found on the continents. In the Atlantic, the Pacific, and in other oceans around the world, a system of ridges, called the mid-ocean ridges, is present. These underwater mountain ranges, shown in **Figure 5,** stretch along the center of much of Earth's ocean floor. This discovery raised the curiosity of many scientists. What formed these mid-ocean ridges?

✔ **Reading Check** *How were mid-ocean ridges discovered?*

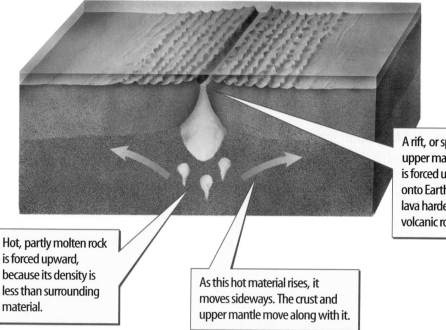

A rift, or split, in the crust and upper mantle forms. Molten rock is forced up into the rift and flows onto Earth's surface as lava. The lava hardens to form new volcanic rock.

Hot, partly molten rock is forced upward, because its density is less than surrounding material.

As this hot material rises, it moves sideways. The crust and upper mantle move along with it.

The Seafloor Moves In the early 1960s, Princeton University scientist Harry Hess suggested an explanation. His now-famous theory is known as **seafloor spreading.** Hess proposed that hot, less dense material below Earth's crust rises toward the surface at the mid-ocean ridges. Then, it flows sideways, carrying the seafloor away from the ridge in both directions, as seen in **Figure 5.**

As the seafloor spreads apart, magma moves upward and flows from the cracks. It becomes solid as it cools and forms new seafloor. As new seafloor moves away from the mid-ocean ridge, it cools, contracts, and becomes denser. This denser, colder seafloor sinks, helping to form the ridge. The theory of seafloor spreading was later supported by the following observations.

Reading Check *How does new seafloor form at mid-ocean ridges?*

Figure 6
Many new discoveries have been made on the seafloor. These giant tube worms inhabit areas near hot water vents along mid-ocean ridges.

Evidence for Spreading In 1968, scientists aboard the research ship *Glomar Challenger* began gathering information about the rocks on the seafloor. *Glomar Challenger* was equipped with a drilling rig that allowed scientists to drill into the seafloor to obtain rock samples. They made a remarkable discovery as they studied the ages of the rocks. Scientists found that the youngest rocks are located at the mid-ocean ridges. The ages of the rocks become increasingly older in samples obtained farther from the ridges, adding to the evidence for seafloor spreading.

Using submersibles along mid-ocean ridges, new seafloor features and life-forms also were discovered there, as shown in **Figure 6.** As molten material rises along the ridges, it brings heat and chemicals that support exotic life-forms in deep, ocean water. Among these are giant clams, mussels, and tube worms.

Physics
INTEGRATION

Magnetic Clues Earth's magnetic field has a north and a south pole. Magnetic lines, or directions, of force leave Earth near the south pole and enter Earth near the north pole. During a magnetic reversal, the lines of magnetic force run the opposite way. Scientists have determined that Earth's magnetic field has reversed itself many times in the past. These reversals occur over intervals of thousands or even millions of years. The reversals are recorded in rocks forming along mid-ocean ridges.

Chemistry
INTEGRATION

Find out what the Curie point is and describe in your Science Journal what happens to iron-bearing minerals when they are heated to the Curie point. Explain how this is important to studies of seafloor spreading.

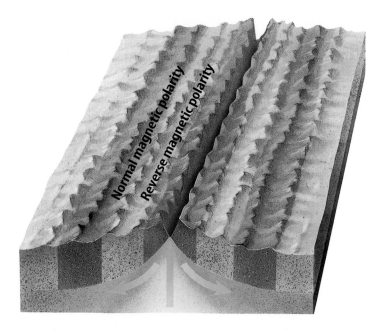

Magnetic Time Scale Iron-bearing minerals, such as magnetite, that are found in the rocks of the seafloor can record Earth's magnetic field direction when they form. Whenever Earth's magnetic field reverses, newly forming iron minerals will record the magnetic reversal.

Using a sensing device called a magnetometer (mag nuh TAH muh tur) to detect magnetic fields, scientists found that rocks on the ocean floor show many periods of magnetic reversal. The magnetic alignment in the rocks reverses back and forth over time in strips parallel to the mid-ocean ridges, as shown in **Figure 7**. A strong magnetic reading is recorded when the polarity of a rock is the same as the polarity of Earth's magnetic field today. Because of this, normal polarities in rocks show up as large peaks. This discovery provided strong support that seafloor spreading was indeed occurring. The magnetic reversals showed that new rock was being formed at the mid-ocean ridges. This helped explain how the crust could move—something that the continental drift hypothesis could not do.

Figure 7
Changes in Earth's magnetic field are preserved in rock that forms on both sides of mid-ocean ridges. *Why is this considered to be evidence of seafloor spreading?*

Section ② Assessment

1. What properties of iron-bearing minerals on the seafloor support the theory of seafloor spreading?

2. How do the ages of the rocks on the ocean floor support the theory of seafloor spreading?

3. How did Harry Hess's hypothesis explain seafloor movement?

4. Why does some partly molten material rise toward Earth's surface?

5. **Think Critically** The ideas of Hess, Wegener, and others emphasize that Earth is a dynamic planet. How is seafloor spreading different from continental drift?

Skill Builder Activities

6. **Concept Mapping** Make a concept map that includes evidence for seafloor spreading using the following phrases: *ages increase away from ridge, pattern of magnetic field reversals, mid-ocean ridge, pattern of ages,* and *reverses back and forth.* **For more help, refer to the** Science Skill Handbook.

7. **Solving Simple Equations** North America is moving about 1.25 cm per year away from a ridge in the middle of the Atlantic Ocean. Using this rate, how much farther apart will North America and the ridge be in 200 million years? **For more help, refer to the** Math Skill Handbook.

Activity

Seafloor Spreading Rates

Ⓗow did scientists use their knowledge of seafloor spreading and magnetic field reversals to reconstruct Pangaea? Try this activity to see how you can determine where a continent may have been located in the past.

What You'll Investigate

Can you use clues, such as magnetic field reversals on Earth, to help reconstruct Pangaea?

Materials

metric ruler
pencil

Goals

■ **Interpret** data about magnetic field reversals. Use these magnetic clues to reconstruct Pangaea.

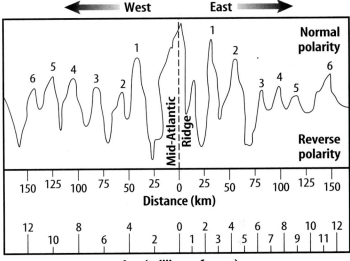

Procedure

1. Study the magnetic field graph above. You will be working only with normal polarity readings, which are the peaks above the baseline in the top half of the graph.

2. Place the long edge of a ruler vertically on the graph. Slide the ruler so that it lines up with the center of peak 1 west of the Mid-Atlantic Ridge.

3. **Determine** and record the distance and age that line up with the center of peak 1 west. Repeat this process for peak 1 east of the ridge.

4. **Calculate** the average age and distance for this pair of peaks.

5. Repeat steps 2 through 4 for the remaining pairs of normal-polarity peaks.

6. **Calculate** the rate of movement in cm per year for the six pairs of peaks. Use the formula rate = distance/time. Convert kilometers to centimeters. For example, to calculate a rate using normal-polarity peak 5, west of the ridge:

$$\text{rate} = \frac{125 \text{ km}}{10 \text{ million years}} = \frac{12.5 \text{ km}}{\text{million years}} = \frac{1,250,000 \text{ cm}}{1,000,000 \text{ years}} = 1.25 \text{ cm/year}$$

Conclude and Apply

1. **Compare** the age of igneous rock found near the mid-ocean ridge with that of igneous rock found farther away from the ridge.

2. If the distance from a point on the coast of Africa to the Mid-Atlantic Ridge is approximately 2,400 km, calculate how long ago that point in Africa was at or near the Mid-Atlantic Ridge.

3. How could you use this method to reconstruct Pangaea?

Theory of Plate Tectonics

As You Read

What You'll Learn

■ **Compare and contrast** different types of plate boundaries.
■ **Explain** how heat inside Earth causes plate tectonics.
■ **Recognize** features caused by plate tectonics.

Vocabulary
plate tectonics
plate
lithosphere
asthenosphere
convection current

Why It's Important
Plate tectonics explains how many of Earth's features form.

Plate Tectonics

The idea of seafloor spreading showed that more than just continents were moving, as Wegener had thought. It was now clear to scientists that sections of the seafloor and continents move in relation to one another.

Plate Movements In the 1960s, scientists developed a new theory that combined continental drift and seafloor spreading. According to the theory of **plate tectonics,** Earth's crust and part of the upper mantle are broken into sections. These sections, called **plates,** move on a plasticlike layer of the mantle. The plates can be thought of as rafts that float and move on this layer.

Composition of Earth's Plates Plates are made of the crust and a part of the upper mantle, as shown in **Figure 8.** These two parts combined are the **lithosphere** (LIH thuh sfihr). This rigid layer is about 100 km thick and generally is less dense than material underneath. The plasticlike layer below the lithosphere is called the **asthenosphere** (as THE nuh sfihr). The rigid plates of the lithosphere float and move around on the asthenosphere.

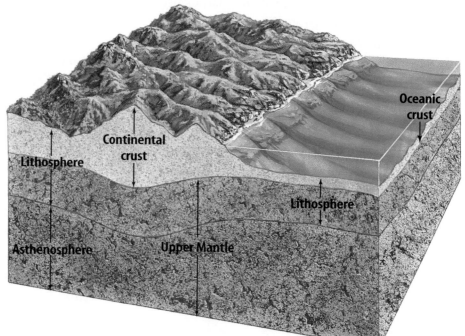

Figure 8
Plates of the lithosphere are composed of oceanic crust, continental crust, and rigid upper mantle.

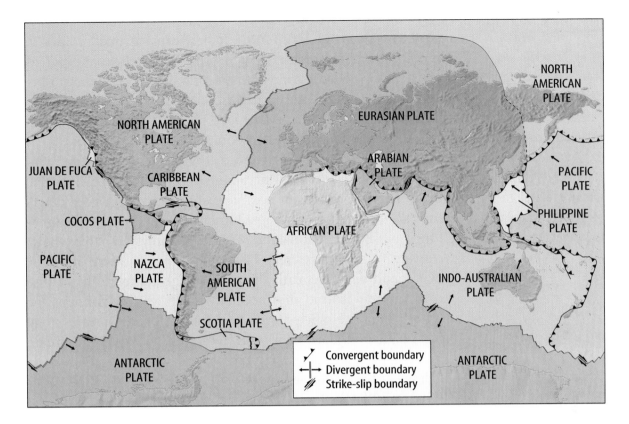

Figure 9
This diagram shows the major plates of the lithosphere, their direction of movement, and the type of boundary between them. *Based on what is shown in this figure, what is happening where the Nazca Plate meets the Pacific Plate?*

Map labels: NORTH AMERICAN PLATE, EURASIAN PLATE, NORTH AMERICAN PLATE, JUAN DE FUCA PLATE, CARIBBEAN PLATE, ARABIAN PLATE, PACIFIC PLATE, COCOS PLATE, AFRICAN PLATE, PHILIPPINE PLATE, PACIFIC PLATE, NAZCA PLATE, SOUTH AMERICAN PLATE, INDO-AUSTRALIAN PLATE, SCOTIA PLATE, ANTARCTIC PLATE, ANTARCTIC PLATE

Legend:
- Convergent boundary
- Divergent boundary
- Strike-slip boundary

Plate Boundaries

When plates move, they can interact in several ways. They can move toward each other and converge, or collide. They also can pull apart or slide alongside one another. When the plates interact, the result of their movement is seen at the plate boundaries, as in **Figure 9.**

Reading Check *What are the general ways that plates interact?*

Movement along any plate boundary means that changes must happen at other boundaries. What is happening to the Atlantic Ocean floor between the North American and African Plates? Compare this with what is happening along the western margin of South America.

Plates Moving Apart The boundary between two plates that are moving apart is called a divergent boundary. You learned about divergent boundaries when you read about seafloor spreading. In the Atlantic Ocean, the North American Plate is moving away from the Eurasian and the African Plates, as shown in **Figure 9.** That divergent boundary is called the Mid-Atlantic Ridge. The Great Rift Valley in eastern Africa might become a divergent plate boundary. There, a valley has formed where a continental plate is being pulled apart. **Figure 10** shows a side view of what a rift valley might look like and illustrates how the hot material rises up where plates separate.

Plates Moving Together If new crust is being added at one location, why doesn't Earth's surface keep expanding? As new crust is added in one place, it disappears below the surface at another. The disappearance of crust can occur when seafloor cools, becomes denser, and sinks. This occurs where two plates move together at a convergent boundary.

When an oceanic plate converges with a less dense continental plate, the denser oceanic plate sinks under the continental plate. The area where an oceanic plate subducts, or goes down, into the mantle is called a subduction zone. Some volcanoes form above subduction zones. **Figure 10** shows how this type of convergent boundary creates a deep-sea trench where one plate bends and sinks beneath the other. High temperatures cause rock to melt around the subducting slab as it goes under the other plate. The newly formed magma is forced upward along these plate boundaries, forming volcanoes. The Andes mountain range of South America contains many volcanoes. They were formed at the convergent boundary of the Nazca and the South American Plates.

Problem-Solving Activity

How well do the continents fit together?

Recall the Explore Activity you performed at the beginning of this chapter. While you were trying to fit pieces of a cut-up photograph together, what clues did you use?

Identifying the Problem

Take a copy of a map of the world and cut out each continent. Lay them on a tabletop and try to fit them together, using techniques you used in the Explore Activity. You will find that the pieces of your Earth puzzle—the continents—do not fit together well. Yet, several of the areas on some continents fit together extremely well.

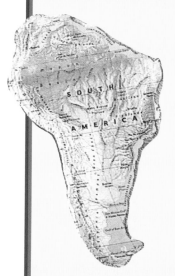

Take out another world map—one that shows the continental shelves as well as the continents. Copy it and cut out the continents, this time including the continental shelves.

Solving the Problem

1. Does including the continental shelves solve the problem of fitting the continents together?
2. Why should continental shelves be included with maps of the continents?

Figure 10

By diverging at some boundaries and converging at others, Earth's plates are continually—but gradually—reshaping the landscape around you. The Mid-Atlantic Ridge, for example, was formed when the North and South American Plates pulled apart from the Eurasian and African Plates (see globe). Some features that occur along plate boundaries— rift valleys, volcanoes, and mountain ranges—are shown on the right and below.

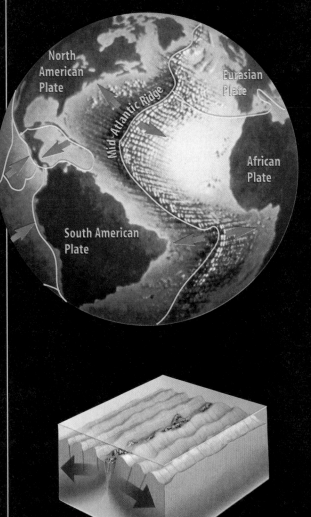

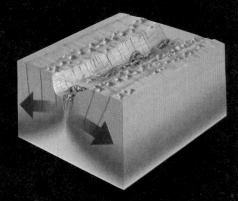

A RIFT VALLEY When continental plates pull apart, they can form rift valleys. The African continent is separating now along the East African Rift Valley.

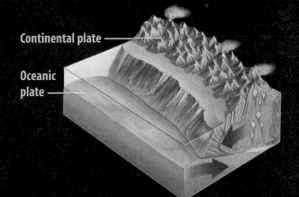

Continental plate

Oceanic plate

SUBDUCTION Where oceanic and continental plates collide, the oceanic plate plunges beneath the less dense continental plate. As the plate descends, molten rock (yellow) forms and rises toward the surface, creating volcanoes.

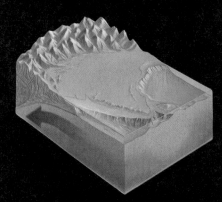

SEA-FLOOR SPREADING A mid-ocean ridge, like the Mid-Atlantic Ridge, forms where oceanic plates continue to separate. As rising magma (yellow) cools, it forms new oceanic crust.

CONTINENTAL COLLISION Where two continental plates collide, they push up the crust to form mountain ranges such as the Himalaya.

Where Plates Collide

A subduction zone also can form where two oceanic plates converge. In this case, the colder, older, denser oceanic plate bends and sinks down into the mantle. The Mariana Islands in the western Pacific are a chain of volcanic islands formed where two oceanic plates collide.

Usually, no subduction occurs when two continental plates collide, as shown in **Figure 10.** Because both of these plates are less dense than the material in the asthenosphere, the two plates collide and crumple up, forming mountain ranges. Earthquakes are common at these convergent boundaries. However, volcanoes do not form because there is no, or little, subduction. The Himalaya in Asia are forming where the Indo-Australian Plate collides with the Eurasian Plate.

Where Plates Slide Past Each Other

The third type of plate boundary is called a transform boundary. Transform boundaries occur where two plates slide past one another. They move in opposite directions or in the same direction at different rates. When one plate slips past another suddenly, earthquakes occur. The Pacific Plate is sliding past the North American Plate, forming the famous San Andreas Fault in California, as seen in **Figure 11.** The San Andreas Fault is part of a transform plate boundary. It has been the site of many earthquakes.

Figure 11
The San Andreas Fault in California occurs along the transform plate boundary where the Pacific Plate is sliding past the North American Plate.

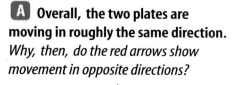 **A** Overall, the two plates are moving in roughly the same direction. *Why, then, do the red arrows show movement in opposite directions?*

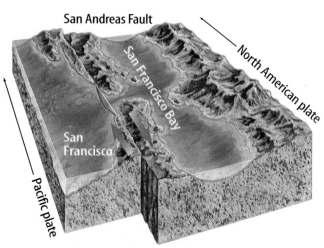

B This photograph shows an aerial view of the San Andreas Fault.

Causes of Plate Tectonics

Many new discoveries have been made about Earth's crust since Wegener's day, but one question still remains. What causes the plates to move? Scientists now think they have a good idea. They think that plates move by the same basic process that occurs when you heat soup.

Convection Inside Earth Soup that is cooking in a pan on the stove contains currents caused by an unequal distribution of heat in the pan. Hot, less dense soup is forced upward by the surrounding, cooler soup. As the hot soup reaches the surface, it cools and sinks back down into the pan. This entire cycle of heating, rising, cooling, and sinking is called a **convection current.** A version of this same process, occurring in the mantle, is thought to be the force behind plate tectonics. Scientists suggest that differences in density cause hot, plasticlike rock to be forced upward toward the surface.

Moving Mantle Material Wegener wasn't able to come up with an explanation for why plates move. Today, researchers who study the movement of heat in Earth's interior have proposed several possible explanations. All of the hypotheses use convection in one way or another. It is, therefore, the transfer of heat inside Earth that provides the energy to move plates and causes many of Earth's surface features. One hypothesis is shown in **Figure 12.** It relates plate motion directly to the movement of convection currents. According to this hypothesis, convection currents cause the movements of plates.

Mini LAB

Modeling Convection Currents

Procedure
1. Pour **water** into a **clear, colorless casserole dish** until it is 5 cm from the top.
2. Center the dish on a **hot plate** and heat it. **WARNING:** *Wear **thermal mitts** to protect your hands.*
3. Add a few drops of **food coloring** to the water above the center of the hot plate.
4. Looking from the side of the dish, observe what happens in the water.
5. Illustrate your observations in your **Science Journal.**

Analysis
1. Determine whether any currents form in the water.
2. Infer what causes the currents to form.

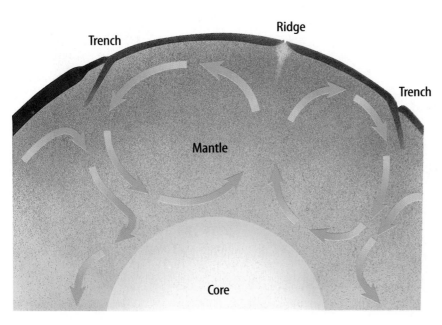

Figure 12
In one hypothesis, convection currents occur throughout the mantle. Such convection currents (see arrows) are the driving force of plate tectonics.

Features Caused by Plate Tectonics

Earth is a dynamic planet with a hot interior. This heat leads to convection, which powers the movement of plates. As the plates move, they interact. The interaction of plates produces forces that build mountains, create ocean basins, and cause volcanoes. When rocks in Earth's crust break and move, energy is released in the form of seismic waves. Humans feel this release as earthquakes. You can see some of the effects of plate tectonics in mountainous regions, where volcanoes erupt, or where landscapes have changed from past earthquake or volcanic activity.

✔ Reading Check *What happens when seismic energy is released as rocks in Earth's crust break and move?*

Normal Faults and Rift Valleys Tension forces, which are forces that pull apart, can stretch Earth's crust. This causes large blocks of crust to break and tilt or slide down the broken surfaces of crust. When rocks break and move along surfaces, a fault forms. Faults interrupt rock layers by moving them out of place. Entire mountain ranges can form in the process, called fault-block mountains, as shown in **Figure 13.** Generally, the faults that form from pull-apart forces are normal faults—faults in which the rock layers above the fault move down when compared with rock layers below the fault.

Rift valleys and mid-ocean ridges can form where Earth's crust separates. Examples of rift valleys are the Great Rift Valley in Africa, and the valleys that occur in the middle of mid-ocean ridges. Examples of mid-ocean ridges include the Mid-Atlantic Ridge and the East Pacific Rise.

Figure 13
Fault-block mountains can form when Earth's crust is stretched by tectonic forces. The arrows indicate the directions of moving blocks. *What type of force occurs when Earth's crust is pulled in opposite directions?*

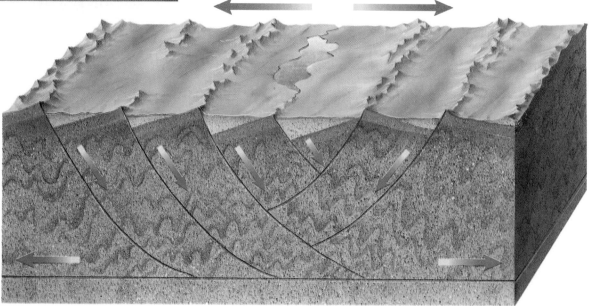

Mountains and Volcanoes Compression forces squeeze objects together. Where plates come together, compression forces produce several effects. As continental plates collide, the forces that are generated cause massive folding and faulting of rock layers into mountain ranges such as the Himalaya, shown in **Figure 14,** or the Appalachian Mountains. The type of faulting produced is generally reverse faulting. Along a reverse fault, the rock layers above the fault surface move up relative to the rock layers below the fault.

Reading Check *What features occur where plates converge?*

As you learned earlier, when two oceanic plates converge, the denser plate is forced beneath the other plate. Curved chains of volcanic islands called island arcs form above the sinking plate. If an oceanic plate converges with a continental plate, the denser oceanic plate slides under the continental plate. Folding and faulting at the continental plate margin can thicken the continental crust to produce mountain ranges. Volcanoes also typically are formed at this type of convergent boundary.

Are there any features caused by plate tectonics in your area? To find out more about these features, see the **Field Guide to Faults and Folds** at the back of the book.

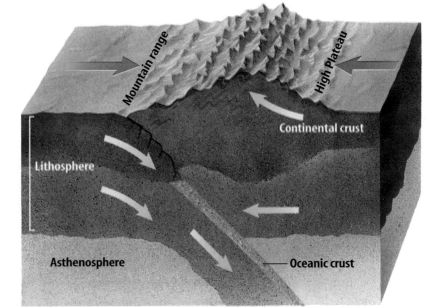

Figure 14
The Himalaya still are forming today as the Indo-Australian Plate collides with the Eurasian Plate.

Figure 15
Most of the movement along a strike-slip fault is parallel to Earth's surface. When movement occurs, human-built structures along a strike-slip fault are offset, as shown here in this road.

Strike-Slip Faults At transform boundaries, two plates slide past one another without converging or diverging. The plates stick and then slide, mostly in a horizontal direction, along large strike-slip faults. In a strike-slip fault, rocks on opposite sides of the fault move in opposite directions, or in the same direction at different rates. This type of fault movement is shown in **Figure 15.** One such example is the San Andreas Fault. When plates move suddenly, vibrations are generated inside Earth that are felt as an earthquake.

Earthquakes, volcanoes, and mountain ranges are evidence of plate motion. Plate tectonics explains how activity inside Earth can affect Earth's crust differently in different locations. You've seen how plates have moved since Pangaea separated. Is it possible to measure how far plates move each year?

Testing for Plate Tectonics

Until recently, the only tests scientists could use to check for plate movement were indirect. They could study the magnetic characteristics of rocks on the seafloor. They could study volcanoes and earthquakes. These methods supported the theory that the plates have moved and still are moving. However, they did not provide proof—only support—of the idea.

New methods had to be discovered to be able to measure the small amounts of movement of Earth's plates. One method, shown in **Figure 16,** uses lasers and a satellite. Now, scientists can measure exact movements of Earth's plates of as little as 1 cm per year.

Physics
INTEGRATION

In which directions do forces act at convergent, divergent, and transform boundaries? Demonstrate these forces using wooden blocks or your hands.

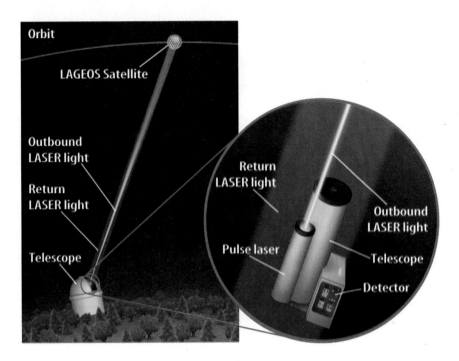

Figure 16
When using the Satellite Laser Ranging System, scientists on the ground aim laser pulses at a satellite. The pulses reflect off the satellite and are used to determine a precise location on the ground.

Current Data Satellite data show that Hawaii is moving toward Japan at a rate of about 8.3 cm per year. Maryland is moving away from England at a rate of 1.7 cm per year. Using such methods, scientists have observed that the plates move at rates ranging from about 1 cm to 12 cm per year.

Section  Assessment

1. What happens to plates at a transform plate boundary?

2. What occurs at plate boundaries that are associated with seafloor spreading?

3. Describe three types of plate boundaries where volcanic eruptions can occur.

4. How are convection currents related to plate tectonics?

5. **Think Critically** Using **Figure 9** and a world map, determine what natural disasters might occur in Iceland. Also determine what disasters might occur in Tibet. Explain why some Icelandic disasters are not expected to occur in Tibet.

Skill Builder Activities

6. **Predicting** Plate tectonic activity causes many events that can be dangerous to humans. One of these events is a seismic sea wave, or tsunami. Learn how scientists predict the arrival time of a tsunami in a coastal area. **For more help, refer to the** Science Skill Handbook.

7. **Using a Word Processor** Write three separate descriptions of the three basic types of plate boundaries—divergent boundaries, convergent boundaries, and transform boundaries. Then draw a sketch of an example of each boundary next to your description. **For more help, refer to the** Technology Skill Handbook.

Predicting Tectonic Activity

The movement of plates on Earth causes forces that build up energy in rocks. The release of this energy can produce vibrations in Earth that you know as earthquakes. Earthquakes occur every day. Many of them are too small to be felt by humans, but each event tells scientists something more about the planet. Active volcanoes can do the same and often form at plate boundaries.

Recognize the Problem

Can you predict tectonically active areas by plotting locations of earthquake epicenters and volcanic eruptions?

Form a Hypothesis

Think about where earthquakes and volcanoes have occurred in the past. Make a hypothesis about whether the locations of earthquake epicenters and active volcanoes can be used to predict tectonically active areas.

Goals

■ **Research** the locations of earthquakes and volcanic eruptions around the world.

■ **Plot** earthquake epicenters and the locations of volcanic eruptions obtained from the Glencoe Science Web site.

■ **Predict** locations that are tectonically active based on a plot of the locations of earthquake epicenters and active volcanoes.

Data Sources

SCIENCE *Online* Go to the Glencoe Science Web site at **science.glencoe.com** for more information about earthquake and volcano sites, hints about earthquake and volcano sites, and data from other students.

Test Your Hypothesis

Plan

1. Make a data table in your Science Journal like the one shown.

2. Collect data for earthquake epicenters and volcanic eruptions for at least the past two weeks. Your data should include the longitude and latitude for each location. For help, refer to the data sources given on the opposite page.

Do

1. Make sure your teacher approves your plan before you start.

2. **Plot** the locations of earthquake epicenters and volcanic eruptions on a map of the world. Use an overlay of tissue paper or plastic.

Locations of Epicenters and Eruptions		
Earthquake Epicenter/ Volcanic Eruption	Longitude	Latitude

3. After you have collected the necessary data, predict where the tectonically active areas on Earth are.

4. **Compare and contrast** the areas that you predicted to be tectonically active with the plate boundary map shown in **Figure 9.**

Analyze Your Data

1. What areas on Earth do you predict to be the locations of tectonic activity?

2. How close did your prediction come to the actual location of tectonically active areas?

Draw Conclusions

1. How could you make your predictions closer to the locations of actual tectonic activity?

2. Would data from a longer period of time help? Explain.

3. What types of plate boundaries were close to your locations of earthquake epicenters? Volcanic eruptions?

4. **Explain** which types of plate boundaries produce volcanic eruptions. Be specific.

Communicating Your Data

SCIENCE Online Find this Internet activity on the Glencoe Science Web site at **science.glencoe .com. Post** your data in the table provided. **Compare** your data to those of other students. Combine your data with those of other students and **plot** these combined data on a map to **recognize** the relationship between plate boundaries, volcanic eruptions, and earthquake epicenters.

Listening In
by Gordon Judge

I'm just a bit of seafloor on this mighty solid sphere.
With no mind to be broadened, I'm quite content down here.
The mantle churns below me, and the sea's in turmoil, too;
But nothing much disturbs me, I'm rock solid through and
  through.

I do pick up occasional low-frequency vibrations –
(I think, although I can't be sure, they're sperm whales'
  conversations).
I know I shouldn't listen in, but what else can I do?
It seems they are all studying for degrees from the OU.

They've mentioned me in passing, as their minds begin improving:
I think I've heard them say "The theory says the sea-floor's
  moving…".
Well, that shook me, I can tell you; yes, it gave me quite a fright.
Yet I've not moved for ages, so I *know* it can't be right.

They call it "Plate Tectonics", this new theory in their noddle.
If they would only ask me, I could tell them it's all twaddle.
Apparently, I "oozed out from a mid-Atlantic split,
Solidified and cooled right down, then moved out bit by bit".

But, how can I be moving, when I know full well myself
That I'm quite firmly anchored to a continental shelf?
"Well, the continent is moving, too; you're *pushing* it, you see,"
I hear those OU whales intone, hydro-acoustically.

Now, my best mate's a sea floor in the mighty East Pacific.
He reckons life is balmy there: the summers are terrific!
He's heard the whale-talk, too, and found it pretty scary.
"Subduction" was the word he heard, which sounded rather hairy.

It was to be his fate, they claimed with undisguised great relish:
A hot and fiery end to things – it really would be hellish.
In fact, he'd end up underneath *my* continent, lengthwise,
So I would be the one to blame for my poor mate's demise.

Well, thank you very much, OU. You've upset my composure.
Next time you send your student whales to look at my exposure
I'll tell them it's a load of tosh: it's *they* who move, not me,
Those arty-smarty blobs of blubber, clogging up the sea!

Understanding Literature

Point of View Point of view refers to the perspective from which an author writes. This poem begins, "I'm just a bit of sea floor…." Right away, you know that the poem, or story, is being told from the point of view of the speaker, or the "first person." Not all first-person stories are told from the point of view of a person. The narrator in this poem is a geological feature, not a person. This point of view helps give the poem a fantastic or outlandish quality. It also gives a playful tone to the poem. What other effects does the first-person narration have on the story?

Science Connection Volcanoes can occur where two plates move toward each other. In the poem, the author gives several clues that a volcano will form. First, the narrator's "best mate" is a seafloor in the Pacific Ocean. When an oceanic plate and a continental plate collide, a volcano will form. The narrator also hears the word *subduction* spoken. Subduction zones occur when one plate sinks under another plate. Rocks melt in the zones where these plates converge, causing magma to move upward and form volcanic mountains. What other clues does the author give that a volcano will form?

Linking Science and Writing

Using Point of View Using the first-person point of view, write an account from the point of view of a living or nonliving thing. You could write an account of an object, such as a pencil, that you use or encounter every day. You also could write from the point of view of a living thing, such as a family pet. Be sure to use the personal pronoun "I" in your account.

Career Connection

Volcanologist

Ed Klimasauskas is a volcanologist at the Cascades Volcano Observatory in Washington State. His job is to study volcanoes in order to predict eruptions. Volcanologists' predictions can save lives: people can be evacuated from danger areas before an eruption occurs. Klimasauskas also educates the public about the hazards of volcanic eruptions and tells people who live near active volcanoes what they can do to be safe in case a volcano erupts. Volcanologists travel all over the world to study new sites.

SCIENCE *Online* To learn more about careers in volcanology, visit the Glencoe Science Web site at **science.glencoe.com**.

Chapter 4 Study Guide

Reviewing Main Ideas

Section 1 Continental Drift

1. Alfred Wegener suggested that the continents were joined together at some point in the past in a large landmass he called Pangaea. Wegener proposed that continents have moved slowly, over millions of years, to their current locations.

2. The puzzlelike fit of the continents, fossils, climatic evidence, and similar rock structures support Wegener's idea of continental drift. However, Wegener could not explain what process could cause the movement of the landmasses. *How do fossils support the hypothesis of continental drift?*

Section 2 Seafloor Spreading

1. Detailed mapping of the ocean floor in the 1950s showed underwater mountains and rift valleys.

2. In the 1960s, Harry Hess suggested seafloor spreading as an explanation for the formation of mid-ocean ridges. *How is magnetic evidence preserved in rocks forming along a mid-ocean ridge?*

■ Normal magnetic polarity
☐ Reversed magnetic polarity
- - - Mid-ocean ridge

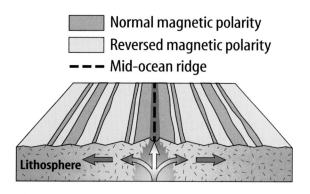

Lithosphere

3. The theory of seafloor spreading is supported by magnetic evidence in rocks and by the ages of rocks on the ocean floor.

Section 3 Theory of Plate Tectonics

1. In the 1960s, scientists combined the ideas of continental drift and seafloor spreading to develop the theory of plate tectonics. The theory states that the surface of Earth is broken into sections called plates that move around on the asthenosphere.

2. Currents in Earth's mantle called convection currents transfer heat in Earth's interior. It is thought that this transfer of heat energy moves plates.

3. Earth is a dynamic planet. As the plates move, they interact, resulting in many of the features of Earth's surface. *How do converging plates form mountains?*

FOLDABLES
Reading & Study Skills

After You Read

To help you review what you learned about plate tectonics, use the Foldable you made at the beginning of the chapter.

Visualizing Main Ideas

Complete the concept map below about continental drift, seafloor spreading, and plate tectonics.

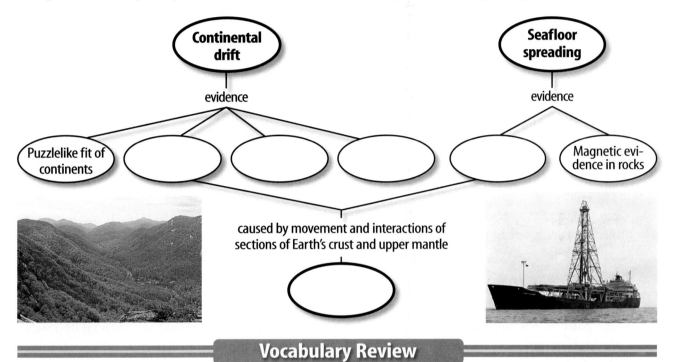

Vocabulary Review

Vocabulary Words

a. asthenosphere
b. continental drift
c. convection current
d. lithosphere
e. Pangaea
f. plate
g. plate tectonics
h. seafloor spreading

Using Vocabulary

Each phrase below describes a vocabulary term from the list. Write the term that matches the phrase describing it.

1. plasticlike layer below the lithosphere

Study Tip

Make a note of anything you don't understand so that you'll remember to ask your teacher about it.

2. idea that continents move slowly across Earth's surface

3. large, ancient landmass that consisted of all the continents on Earth

4. process that forms new seafloor as hot material is forced upward

5. driving force for plate movement

6. composed of oceanic or continental crust and upper mantle

7. explains locations of mountains, trenches, and volcanoes

8. piece of the lithosphere that moves over a plasticlike layer

9. theory proposed by Harry Hess that includes processes along mid-ocean ridges

10. forms as warm material rises and cold material sinks

Checking Concepts

Choose the word or phrase that best answers the question.

1. Which layer of Earth contains the asthenosphere?
   A) crust   C) outer core
   B) mantle   D) inner core

2. What type of plate boundary is the San Andreas Fault part of?
   A) divergent   C) convergent
   B) subduction   D) transform

3. What hypothesis states that continents slowly moved to their present positions on Earth?
   A) subduction   C) continental drift
   B) seafloor spreading D) erosion

4. Which plate is subducting beneath the South American Plate to form the Andes mountain range?
   A) North American   C) Indo-Australian
   B) African   D) Nazca

5. Which of the following features indicates that many continents were once near Earth's south pole?
   A) glacial deposits   C) volcanoes
   B) mid-ocean ridges   D) earthquakes

6. What evidence in rocks supports the theory of seafloor spreading?
   A) plate movement   C) subduction
   B) magnetic reversals D) convergence

7. Which type of plate boundary is the Mid-Atlantic Ridge a part of?
   A) convergent   C) transform
   B) divergent   D) lithosphere

8. What theory states that plates move around on the asthenosphere?
   A) continental drift   C) subduction
   B) seafloor spreading   D) plate tectonics

9. What forms when one plate slides past another plate?
   A) transform boundary
   B) divergent boundary
   C) subduction zone
   D) mid-ocean ridge

10. When oceanic plates collide, what volcanic landforms are made?
    A) folded mountains
    B) island arcs
    C) strike-slip faults
    D) mid-ocean ridges

Thinking Critically

11. Why do many earthquakes but few volcanic eruptions occur in the Himalaya?

12. Glacial deposits often form at high latitudes near the poles. Explain why glacial deposits have been found in Africa.

13. How is magnetism used to support the theory of seafloor spreading?

14. Explain why volcanoes do not form along the San Andreas Fault.

15. Explain why the fossil of an ocean fish found on two different continents would not be good evidence of continental drift.

Developing Skills

16. **Forming Hypotheses** Mount St. Helens in the Cascade Range is a volcano. Use **Figure 9** and a U.S. map to hypothesize how it might have formed.

17. **Measuring in SI** Movement along the African Rift Valley is about 2.1 cm per year. If plates continue to move apart at this rate, how much larger will the rift be (in meters) in 1,000 years? In 15,500 years?

18. Concept Mapping Make an events chain concept map that describes seafloor spreading along a divergent plate boundary. Choose from the following phrases: *magma cools to form new seafloor, convection currents circulate hot material along divergent boundary,* and *older seafloor is forced apart.*

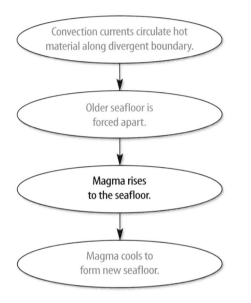

Convection currents circulate hot material along divergent boundary.

Older seafloor is forced apart.

Magma rises to the seafloor.

Magma cools to form new seafloor.

Performance Assessment

19. Observe and Infer In the MiniLab Modeling Convection Currents, you observed convection currents produced in water as it was heated. Repeat the experiment, placing sequins, pieces of wood, or pieces of rubber bands into the water. How do their movements support your observations and inferences from the MiniLab?

TECHNOLOGY

Go to the Glencoe Science Web site at **science.glencoe.com** or use the **Glencoe Science CD-ROM** for additional chapter assessment.

THE PRINCETON REVIEW **Test Practice**

Ms. Fernandez was leading a class discussion on plate tectonics and Earth's interior.

Study the diagram below and then answer the following questions.

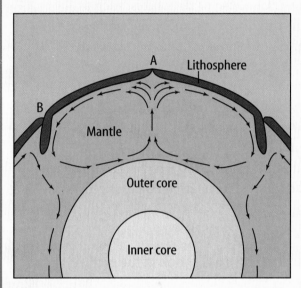

1. Suppose that the arrows in the diagram represent patterns of convection in Earth's mantle. Which type of plate boundary is most likely to form along the region labeled "A"?
   A) transform
   B) reverse
   C) convergent
   D) divergent

2. Which statement is true of the region marked "B" on the diagram?
   F) Plates separate and slip past one another sideways.
   G) Plates diverge and volcanoes form.
   H) Plates converge and volcanoes form.
   J) Plates collapse and form a strike-slip boundary.

Earthquakes

More than 20,000 deaths and at least 166,000 injuries resulted from a powerful earthquake in India on January 26, 2001. Collapse of structures, such as this building in Ahmedabad, India, is among the greatest dangers associated with earthquakes. What causes earthquakes? Why do some areas experience repeated earthquakes while other areas rarely do? In this chapter you'll learn the answers to these and other questions. You'll also learn how earthquake damage can be reduced.

What do you think?

Science Journal Look at the picture below with a classmate. Discuss what you think this might be. Here's a hint: *Something must have been shaking to make these waves.* Write your answer or best guess in your Science Journal.

Why do earthquakes occur? The bedrock beneath the soil can break and form cracks known as faults. When blocks of rock move past each other along a fault, they cause the ground to shake. Why don't rocks move all the time, causing constant earthquakes? You'll find out during this activity.

Model stress buildup along faults

1. Tape a sheet of medium-grain sandpaper to the tabletop.
2. Tape a second sheet of sandpaper to the book cover on a textbook.
3. Place the book on the table so that both sheets of sandpaper meet.
4. Tie two large, thick rubber bands together and loop one of the rubber bands around the edge of the book so that it is not touching the sandpaper.
5. Pull on the free rubber band until the book moves and observe this movement.

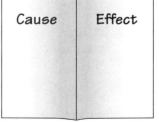

Observe
Write a paragraph in your Science Journal describing how the book moved and explaining how this activity modeled the buildup of stress along a fault.

Before You Read

Making a Cause and Effect Study Fold Make the following Foldable to help you understand the cause and effect relationship of earthquakes and Earth's crust.

1. Place a sheet of paper in front of you so the long side is at the top. Fold the paper in half from the left side to the right side and then unfold.
2. Label the left side of the paper *Cause* and the right side *Effect*. Refold the paper.
3. Before you read the chapter, draw a cross section of Earth's crust showing what you think happens during an earthquake on the outside of your Foldable.
4. As you read the chapter, change your drawing and list causes of earthquakes on the inside of your Foldable.

Forces Inside Earth

What You'll Learn

- **Explain** how earthquakes result from the buildup of energy in rocks.
- **Describe** how compression, tension, and shear forces make rocks move along faults.
- **Distinguish** among normal, reverse, and strike-slip faults.

Vocabulary

fault
earthquake
normal fault

reverse fault
strike-slip fault

Why It's Important

Earthquakes are among the most dramatic of all natural disasters on Earth.

Earthquake Causes

Recall the last time you used a rubber band. Rubber bands stretch when you pull them. Because they are elastic, they return to their original shape once the force is released. However, if you stretch a rubber band too far, it will break. A wooden craft stick behaves in a similar way. When a force is first applied to the stick, it will bend and change shape, as shown in **Figure 1A.** The energy needed to bend the stick is stored inside the stick as potential energy. If the force keeping the stick bent is removed, the stick will return to its original shape, and the stored energy will be released as energy of motion.

Fault Formation There is a limit to how far a wooden craft stick can bend. This is called its elastic limit. Once its elastic limit is passed, the stick breaks, as shown in **Figure 1B.** Rocks behave in a similar way. Up to a point, applied forces cause rocks to bend and stretch, undergoing what is called elastic deformation. Once the elastic limit is passed, the rocks may break. When rocks break, they move along surfaces called **faults.** A tremendous amount of force is required to overcome the strength of rocks and to cause movement along a fault. Rock along one side of a fault can move up, down, or sideways in relation to rock along the other side of the fault.

A   **B**

Figure 1
The **A** bending and **B** breaking of wooden craft sticks are similar to how rocks bend and break.

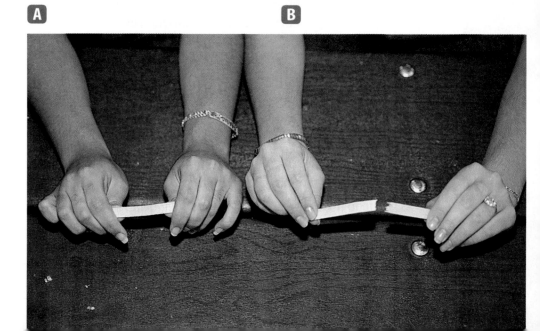

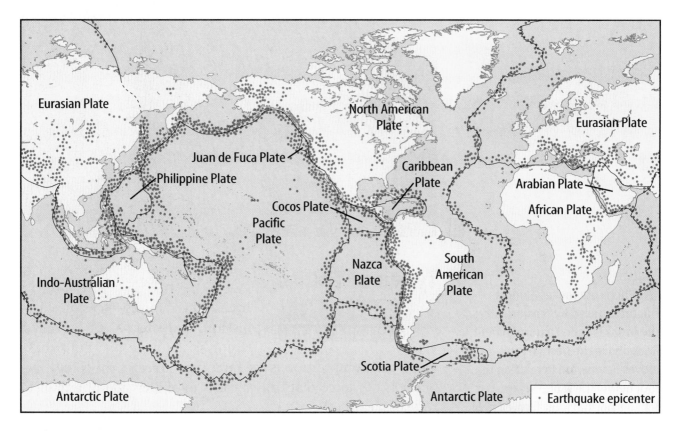

Eurasian Plate

North American Plate

Juan de Fuca Plate

Philippine Plate

Caribbean Plate

Eurasian Plate

Arabian Plate

African Plate

Cocos Plate

Pacific Plate

Nazca Plate

South American Plate

Indo-Australian Plate

Scotia Plate

Antarctic Plate

Antarctic Plate

· Earthquake epicenter

What causes faults? What produces the forces that cause rocks to break and faults to form? The surface of Earth is in constant motion because of forces inside the planet. These forces cause sections of Earth's surface, called plates, to move. This movement puts stress on the rocks near the plate edges. To relieve this stress, the rocks tend to bend, compress, or stretch. If the force is great enough, the rocks will break. An **earthquake** is the vibrations produced by the breaking of rock. **Figure 2** shows how the locations of earthquakes outline the plates that make up Earth's surface.

Reading Check *Why do most earthquakes occur near plate boundaries?*

How Earthquakes Occur As rocks move past each other along a fault, their rough surfaces catch, temporarily halting movement along the fault. However, forces keep driving the rocks to move. This action builds up stress at the points where the rocks are stuck. The stress causes the rocks to bend and change shape. When the rocks are stressed beyond their elastic limit, they break, move along the fault, and return to their original shapes. An earthquake results. Earthquakes range from unnoticeable vibrations to devastating waves of energy. Regardless of their intensity, most earthquakes result from rocks moving over, under, or past each other along fault surfaces.

Figure 2
The dots represent the epicenters of major earthquakes over a ten-year period. Note that most earthquakes occur near plate boundaries. *Why do earthquakes rarely occur in the middle of plates?*

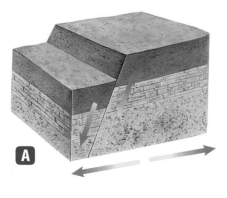

B

Figure 3

A When rock moves along a fracture caused by tension forces, the break is called a normal fault. Rock above the normal fault moves downward in relation to rock below the fault surface.
B This normal fault formed near Kanab, Utah.

Figure 4

A Compression forces in rocks form reverse faults. The rock above the reverse fault surface moves upward in relation to the rock below the fault surface.
B Rock layers have been offset along this reverse fault.

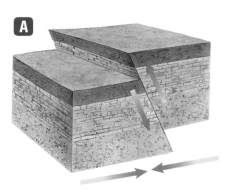

Types of Faults

Three types of forces—tension, compression, and shear—act on rocks. Tension is the force that pulls rocks apart, and compression is the force that squeezes rocks together. Shear is the force that causes rocks on either side of a fault to slide past each other.

Normal Faults Tensional forces inside Earth cause rocks to be pulled apart. When rocks are stretched by these forces, a normal fault can form. Along a **normal fault,** rock above the fault surface moves downward in relation to rock below the fault surface. The motion along a normal fault is shown in **Figure 3A.** Notice the normal fault shown in the photograph in **Figure 3B.**

Reverse Faults Reverse faults result from compression forces that squeeze rock. **Figure 4A** shows the motion along a reverse fault. If rock breaks from forces pushing from opposite directions, rock above a **reverse fault** surface is forced up and over the rock below the fault surface. **Figure 4B** shows a large reverse fault in California.

B

Figure 5

A Shear forces push on rock in opposite—but not directly opposite—horizontal directions. When they are strong enough, these forces split rock and create strike-slip faults. Little vertical movement occurs along a strike-slip fault. **B** The North American Plate and the Pacific Plate slide past each other along the San Andreas Fault, a strike-slip fault, in California.

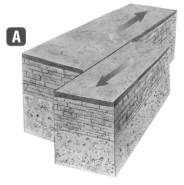

Strike-Slip Faults At a **strike-slip fault,** shown in **Figure 5A,** rocks on either side of the fault are moving past each other without much upward or downward movement. **Figure 5B** shows the largest fault in California—the San Andreas Fault—which stretches more than 1,100 km through the state. The San Andreas Fault is the boundary between two of Earth's plates that are moving sideways past each other.

✔ **Reading Check** *What is a strike-slip fault?*

Section ① Assessment

1. What is an earthquake?

2. The Himalaya in Tibet formed when two of Earth's plates collided. What type of faults would you expect to find in these mountains? Why?

3. In what direction do rocks above a normal fault surface move?

4. Why is California's San Andreas Fault a strike-slip fault?

5. **Think Critically** Why is it easier to predict where an earthquake will occur than it is to predict when it will occur?

Skill Builder Activities

6. **Forming Hypotheses** Hypothesize why the chances of an earthquake occurring along a fault increase rather than decrease as time since the last earthquake passes. **For more help, refer to the** Science Skill Handbook.

7. **Using Graphics Software** Use a graphics program to make models of the three types of faults—normal, reverse, and strike-slip. Add arrows to show the directions of movement along both sides of each type. **For more help, refer to the** Technology Skill Handbook.

Features of Earthquakes

As You Read

What You'll Learn
- **Explain** how earthquake energy travels in seismic waves.
- **Distinguish** among primary, secondary, and surface waves.
- **Describe** the structure of Earth's interior.

Vocabulary
seismic wave
focus
primary wave
secondary wave
surface wave
epicenter
seismograph

Why It's Important
Seismic waves are responsible for most damage caused by earthquakes.

Seismic Waves

When two people hold opposite ends of a rope and shake one end, as shown in **Figure 6,** they send energy through the rope in the form of waves. Like the waves that travel through the rope, **seismic** (SIZE mihk) **waves** generated by an earthquake travel through Earth. During a strong earthquake, the ground moves forward and backward, heaves up and down, and shifts from side to side. The surface of the ground can ripple like waves do in water. Imagine trying to stand on ground that had waves traveling through it. This is what you might experience during a strong earthquake.

Origin of Seismic Waves You learned earlier that rocks move past each other along faults, creating stress at points where the rocks' irregular surfaces catch each other. The stress continues to build up until the elastic limit is exceeded and energy is released in the form of seismic waves. The point where this energy release first occurs is the **focus** (plural, *foci*) of the earthquake. The foci of most earthquakes are within 65 km of Earth's surface. A few have been recorded as deep as 700 km. Seismic waves are produced and travel outward from the earthquake focus.

Figure 6
Some seismic waves are similar to the wave that is traveling through the rope. Note that the rope moves perpendicular to the wave direction.

Primary Waves When earthquakes occur, three different types of seismic waves are produced. All of the waves are generated at the same time, but each behaves differently within Earth. **Primary waves** (P-waves) cause particles in rocks to move back and forth in the same direction that the wave is traveling. If you squeeze one end of a coiled spring and then release it, you cause it to compress and then stretch as the wave travels through the spring, as shown in **Figure 7.** Particles in rocks also compress and then stretch apart, transmitting primary waves through the rock.

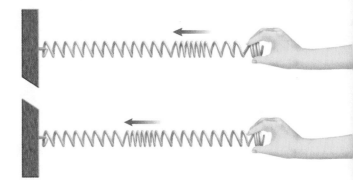

Figure 7
Primary waves move through Earth the same way that a wave travels through a coiled spring.

Secondary and Surface Waves Secondary waves (S-waves) move through Earth by causing particles in rocks to move at right angles to the direction of wave travel. The wave traveling through the rope shown in **Figure 6** is an example of a secondary wave.

Surface waves cause most of the destruction resulting from earthquakes. **Surface waves** move rock particles in a backward, rolling motion and a side-to-side, swaying motion, as shown in **Figure 8.** Many buildings are unable to withstand intense shaking because they are made with stiff materials. The buildings fall apart when surface waves cause different parts of the building to move in different directions.

Physics
INTEGRATION

When sound is produced, waves move through air or some other material. Research sound waves to find out which type of seismic wave they are similar to.

☑ **Reading Check** *Why do surface waves damage buildings?*

Surface waves are produced when earthquake energy reaches the surface of Earth. Surface waves travel outward from the epicenter. The earthquake **epicenter** (EH pi sen tur) is the point on Earth's surface directly above the earthquake focus. Find the focus and epicenter in **Figure 9.**

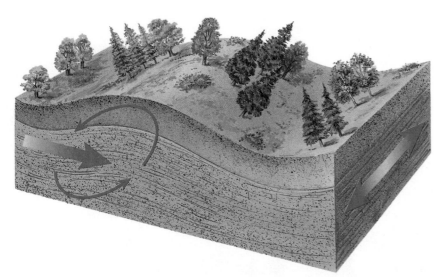

Figure 8
Surface waves move rock particles in a backward, rolling motion and a side-to-side, swaying motion. *How does this movement differ from rock movement caused by secondary waves?*

Figure 9

As the plates that form Earth's lithosphere move, great stress is placed on rocks. They bend, stretch, and compress. Occasionally, rocks break, producing earthquakes that generate seismic waves. As shown here, different kinds of seismic waves—each with distinctive characteristics—move outward from the focus of the earthquake.

C The point on Earth's surface directly above an earthquake's focus is known as the epicenter. Surface waves spread out from the epicenter like ripples in a pond.

D The amplitudes, or heights, of surface waves are greater than those of primary and secondary waves. Surface waves cause the most damage during an earthquake.

B Primary waves and secondary waves originate at the focus and travel outward in all directions. Primary waves travel about twice as fast as secondary waves.

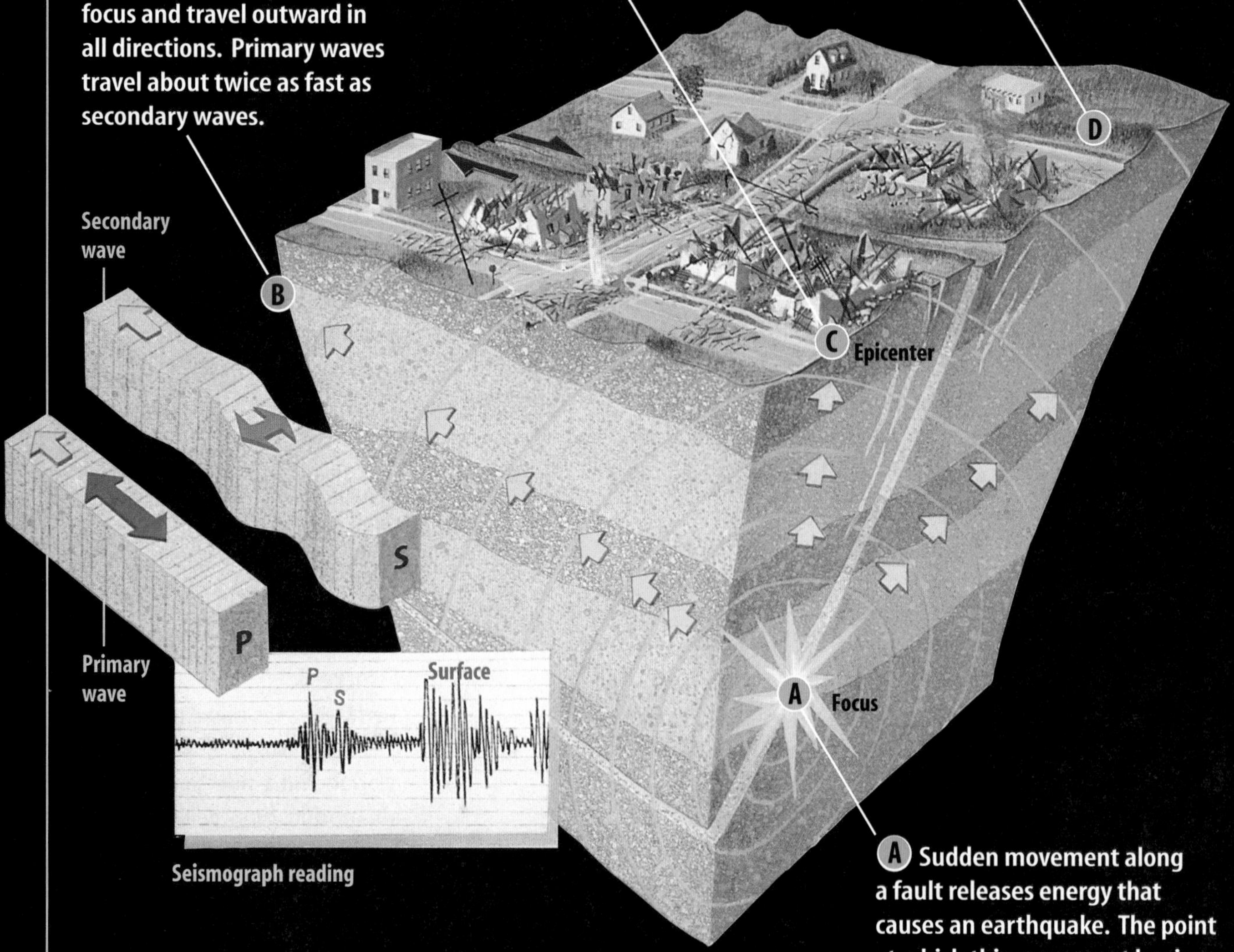

Secondary wave

B

D

Primary wave

S

P

C Epicenter

A Focus

Surface

P S

Seismograph reading

A Sudden movement along a fault releases energy that causes an earthquake. The point at which this movement begins is called the earthquake's focus.

Locating an Epicenter

Different seismic waves travel through Earth at different speeds. Primary waves are the fastest, secondary waves are slower, and surface waves are the slowest. Can you think of a way this information could be used to determine how far away an earthquake epicenter is? Think of the last time you saw two people running in a race. You probably noticed that the faster person got further ahead as the race continued. Like runners in a race, seismic waves travel at different speeds.

Scientists have learned how to use the different speeds of seismic waves to determine the distance to an earthquake epicenter. When an epicenter is far from a location, the primary wave has more time to put distance between it and the secondary and surface waves, just like the fastest runner in a race.

Measuring Seismic Waves Seismic waves from earthquakes are measured with an instrument known as a **seismograph.** Seismographs register the waves and record the time that each arrived. Seismographs consist of a rotating drum of paper and a pendulum with an attached pen. When seismic waves reach the seismograph, the drum vibrates but the pendulum remains at rest. The stationary pen traces a record of the vibrations on the moving drum of paper. The paper record of the seismic event is called a seismogram. **Figure 10** shows two types of seismographs that measure either vertical or horizontal ground movement, depending on the orientation of the drum.

SCIENCE *Online*

Research Visit the Glencoe Science Web site at **science.glencoe.com** to learn about the National Earthquake Information Center and the World Data Center for Seismology. Share what you learn with your class.

Figure 10

Seismographs differ according to whether they are intended to measure horizontal or vertical seismic motions. *Why can't one seismograph measure both horizontal and vertical motions?*

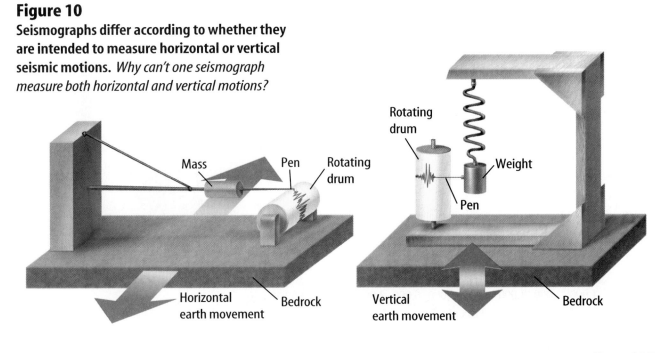

Figure 11
Primary waves arrive at a seismograph station before secondary waves do.

A This graph shows the distance that primary and secondary waves travel over time. By measuring the difference in arrival times, a seismologist can determine the distance to the epicenter.

Seismograph Stations Each type of seismic wave reaches a seismograph station at a different time based on its speed. Primary waves arrive first at seismograph stations, and secondary waves, which travel slower, arrive second, as shown in the graph in **Figure 11A.** Because surface waves travel slowest, they arrive at seismograph stations last.

If seismic waves reach three or more seismograph stations, the location of the epicenter can be determined. To locate an epicenter, scientists draw circles around each station on a map. The radius of each circle equals that station's distance from the earthquake epicenter. The point where all three circles intersect, shown in **Figure 11B,** is the location of the earthquake epicenter.

Seismologists usually describe earthquakes based on their distances from the seismograph. Local events occur less than 100 km away. Regional events occur 100 km to 1,400 km away. Teleseismic events are those that occur at distances greater than 1,400 km.

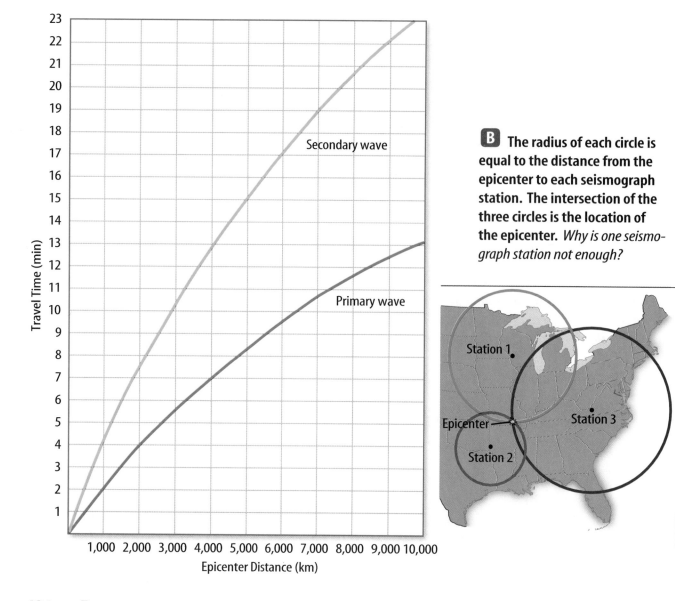

B The radius of each circle is equal to the distance from the epicenter to each seismograph station. The intersection of the three circles is the location of the epicenter. *Why is one seismograph station not enough?*

Basic Structure of Earth

Figure 12 shows Earth's internal structure. At the very center of Earth is a solid, dense inner core made mostly of iron with smaller amounts of nickel, oxygen, silicon, and sulfur. Pressure from the layers above causes the inner core to be solid. Above the solid inner core lies the liquid outer core, which also is made mainly of iron.

✓ Reading Check *How do the inner and outer cores differ?*

Earth's mantle is the largest layer, lying directly above the outer core. It is made mostly of silicon, oxygen, magnesium, and iron. The mantle often is divided into an upper part and a lower part based on changing seismic wave speeds. A portion of the upper mantle, called the asthenosphere (as THE nuh sfihr), consists of weak rock that can flow slowly.

Earth's Crust The outermost layer of Earth is the crust. Together, the crust and a part of the mantle just beneath it make up Earth's lithosphere (LIH thuh sfihr). The lithosphere is broken into a number of plates that move over the asthenosphere beneath it.

The thickness of Earth's crust varies. It is more than 60 km thick in some mountainous regions and less than 5 km thick under some parts of the oceans. Compared to the mantle, the crust contains more silicon and aluminum and less magnesium and iron. Earth's crust generally is less dense than the mantle beneath it.

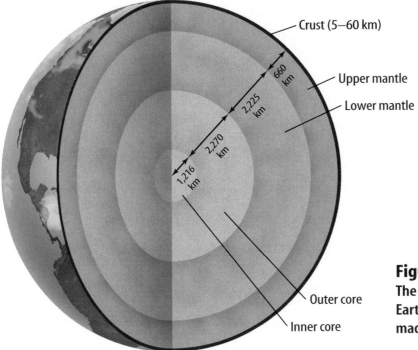

Crust (5–60 km)
Upper mantle
Lower mantle
660 km
2,225 km
2,270 km
1,216 km
Outer core
Inner core

Figure 12
The internal structure of Earth shows that it is made of different layers.

Mini LAB

Interpreting Seismic Wave Data

Procedure
1. Use the **graph** in **Figure 11** to determine the difference in arrival times for primary and secondary waves at the distances listed in the data table below. Two examples are provided for you.

Wave Data	
Distance (km)	**Difference in Arrival Time**
1,500	2 min, 50 s
2,250	
2,750	
3,000	
4,000	5 min, 55 s
7,000	
9,000	

2. Use the graph to determine the differences in arrival times for the other distances in the table.

Analysis
1. What happens to the difference in arrival times as the distance from the earthquake increases?
2. If the difference in arrival times at a seismograph station is 6 min, 30 s, how far away is the epicenter?

Mapping Earth's Internal Structure As shown in **Figure 13,** the speeds and paths of seismic waves change as they travel through materials with different densities. By studying seismic waves that have traveled through Earth, scientists have identified different layers with different densities. In general, the densities increase with depth as pressures increase. Studying seismic waves has allowed scientists to map Earth's internal structure without being there.

Early in the twentieth century, scientists discovered that large areas of Earth don't receive seismic waves from an earthquake. In the area on Earth between 105° and 140° from the earthquake focus, no waves are detected. This area, called the shadow zone, is shown in **Figure 13.** Secondary waves are not transmitted through a liquid, so they stop when they hit the liquid outer core. Primary waves are slowed and bent but not stopped by the liquid outer core. Because of this, scientists concluded that the outer core and mantle are made of different materials. Primary waves speed up again as they travel through the solid inner core. The bending of primary waves and the stopping of secondary waves create the shadow zone.

✔ **Reading Check** *Why do seismic waves change speed as they travel through Earth?*

Figure 13
Seismic waves bend and change speed as the density of rock changes. Primary waves bend when they contact the outer core, and secondary waves are stopped completely. This creates a shadow zone where no seismic waves are received.

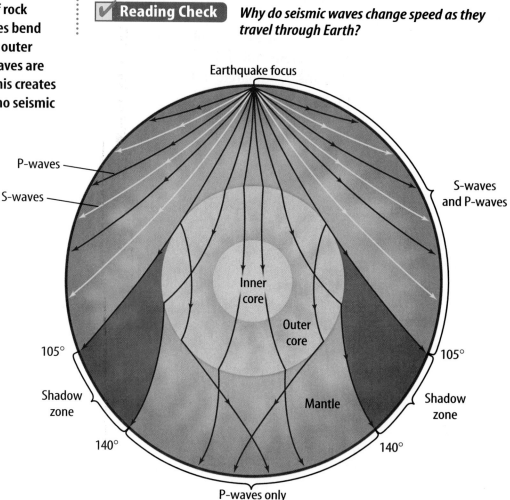

Earthquake focus

P-waves

S-waves

S-waves and P-waves

Inner core

Outer core

105°

105°

Shadow zone

Shadow zone

Mantle

140°

140°

P-waves only

Layer Boundaries **Figure 14** shows how seismic waves change speed as they pass through layers of Earth. Seismic waves speed up when they pass through the bottom of the crust and enter the upper mantle, shown on the far left of the graph. This boundary between the crust and upper mantle is called the Mohorovicic discontinuity (moh huh ROH vee chihch • dis kahn tuh NEW uh tee), or Moho.

The mantle is divided into layers based on changes in seismic wave speeds. For example, primary and sec-

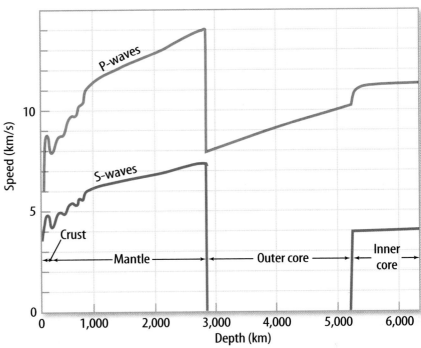

Seismic Wave Speeds

ondary waves slow down again when they reach the astheno- sphere. Then, they generally speed up as they move through a more solid region of the mantle below the asthenosphere.

The core is divided into two layers based on how seismic waves travel through it. Secondary waves do not travel through the liquid outer core, as you can see in the graph. Primary waves slow down when they reach the outer core, but they speed up again upon reaching the solid inner core.

Figure 14
Changes in the speeds of seismic waves allowed scientists to detect boundaries between Earth's layers. S waves in the inner core form when P waves strike its surface.

Section 2 Assessment

1. How many seismograph stations are needed to determine the location of an epicenter? Explain.

2. Name the layers of Earth's interior.

3. What makes up most of Earth's inner core?

4. What are the three types of seismic waves? Which one does the most damage to property?

5. **Think Critically** Why do some seismo- graph stations receive both primary and secondary waves from an earthquake but other stations don't?

Skill Builder Activities

6. **Predicting** What will happen to the distance between two opposite walls of a room as pri- mary waves move through the room? **For more help, refer to the** Science Skill Handbook.

7. **Solving One-Step Equations** Primary waves travel about 6 km/s through Earth's crust. The distance from Los Angeles, California, to Phoenix, Arizona, is about 600 km. How long would it take primary waves to travel between the two cities? **For more help, refer to the** Math Skill Handbook.

Activity

Epicenter Location

In this activity you can plot the distance of seismograph stations from the epicenters of earthquakes and determine the earthquake epicenters.

What You'll Investigate

Can plotting the distance of several seismograph stations from two earthquake epicenters allow you to determine the locations of the two epicenters?

Materials

string   globe
metric ruler   chalk

Goals

- **Plot** the distances from several seismograph stations based on primary and secondary wave arrival times.
- **Interpret** the location of earthquake epicenters from these plots.

Earthquake Data			
Location of Seismograph	**Wave**	**Wave Arrival Times**	
		Earthquake A	**Earthquake B**
New York, New York	P S	2:24:05 P.M. 2:29:15 P.M.	1:19:42 P.M. 1:25:27 P.M.
Seattle, Washington	P S	2:24:40 P.M. 2:30:10 P.M.	1:14:37 P.M. 1:16:57 P.M.
Rio de Janeiro, Brazil	P S	2:29:10 P.M. 2:37:50 P.M.	— —
Paris, France	P S	2:30:30 P.M. 2:40:10 P.M.	1:24:57 P.M. 1:34:27 P.M.
Tokyo, Japan	P S	— —	1:24:27 P.M. 1:33:27 P.M.

Procedure

1. Determine the difference in arrival time between the primary and secondary waves at each station for each earthquake listed in the table.

2. After you determine the arrival time differences for each seismograph station, use the graph in **Figure 11** to determine the distance in kilometers of each seismograph from the epicenter of each earthquake. Record these data in a data table. For example, the difference in arrival times in Paris for earthquake B is 9 min, 30 s. On the graph, the primary and secondary waves are separated along the vertical axis by 9 min, 30 s at a distance of 8,975 km.

3. Using the string, measure the circumference of the globe. Determine a scale of centimeters of string to kilometers on Earth's surface. (Earth's circumference is 40,000 km.)

4. For each earthquake, place one end of the string at each seismic station location on the globe. Use the chalk to draw a circle with a radius equal to the distance to the earthquake's epicenter.

5. **Identify** the epicenter for each earthquake.

Conclude and Apply

1. How is the distance of a seismograph from the earthquake related to the arrival times of the waves?

2. What is the location of the epicenter for each earthquake?

3. How many stations were needed to locate each epicenter accurately?

4. **Explain** why some seismographs didn't receive secondary waves from some quakes.

3 People and Earthquakes

Earthquake Activity

Imagine awakening in the middle of the night with your bed shaking, windows shattering, and furniture crashing together. That's what many people in Northridge, California, experienced at 4:30 A.M. on January 17, 1994. The ground beneath Northridge shook violently—it was an earthquake.

Although the earthquake lasted only 15 s, it killed 51 people, injured more than 9,000 people, and caused $44 billion in damage. More than 22,000 people were left homeless. **Figure 15A** shows some of the damage caused by the Northridge earthquake. **Figure 15B** shows the record of the Northridge earthquake on a seismogram.

Earthquakes are natural geological events that provide information about Earth. Unfortunately, they also cause billions of dollars in property damage and kill an average of 10,000 people every year. With so many lives lost and such destruction, it is important for scientists to learn as much as possible about earthquakes to try to reduce their impact on society.

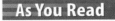
As You Read

What You'll Learn
- **Explain** where most earthquakes in the United States occur.
- **Describe** how scientists measure earthquakes.
- **List** ways to make your classroom and home more earthquake-safe.

Vocabulary
magnitude
liquefaction
tsunami

Why It's Important
Earthquake preparation can save lives and reduce damage.

Figure 15
The 1994 Northridge, California, earthquake was a costly disaster. **A** Several major highways were damaged. **B** A seismograph made this record, called a seismogram, of the earthquake.

Figure 16
The 1999 earthquake in Turkey released about 32 times more energy than the 1994 Northridge earthquake did.

Studying Earthquakes Scientists who study earthquakes and seismic waves are seismologists. As you learned earlier, the instrument that is used to record primary, secondary, and surface waves from earthquakes all over the world is called a seismograph. Seismologists can use records from seismographs, called seismograms, to learn more than just where the epicenter of an earthquake is located.

Measuring Earthquake Magnitude The height of the lines traced on the paper record of a seismograph is a measure of the energy that is released, or the **magnitude,** of the earthquake. The Richter magnitude scale is used to describe the strength of an earthquake and is based on the height of the lines on the seismogram. The Richter scale has no upper limit. However, scientists think that a value of about 9.5 would be the maximum strength an earthquake could register. For each increase of 1.0 on the Richter scale, the height of the line on a seismogram is ten times greater. However, about 32 times as much energy is released for every increase of 1.0 on the scale. For example, an earthquake with a magnitude of 8.5 releases about 32 times more energy than an earthquake with a magnitude of 7.5. **Figure 16** shows damage from the 7.8-magnitude earthquake in Turkey in 1999. **Table 1** is a list of some large-magnitude earthquakes that have occurred around the world and the damage they have caused.

Most of the earthquakes you hear about are large ones that cause great damage. However, of all the earthquakes detected throughout the world each year, most have magnitudes too low to be felt by humans. Scientists record thousands of earthquakes every day with magnitudes of less than 3.0. Each year, about 55,000 earthquakes are felt but cause little or no damage. These minor earthquakes have magnitudes that range from approximately 3.0 to 4.9 on the Richter scale.

Table 1 Large-Magnitude Earthquakes			
Year	**Location**	**Magnitude**	**Deaths**
1556	Shensi, China	?	830,000
1755	Lisbon, Portugal	8.8 (est.)	70,000
1811–12	New Madrid, MO	8.3 (est.)	few
1886	Charleston, SC	?	60
1906	San Francisco, CA	8.3	700 to 800
1923	Tokyo, Japan	9.2	143,000
1960	Chile	9.5	490 to 2,290
1964	Prince William Sound, AK	8.5	131
1976	Tangshan, China	8.2	242,000
1990	Iran	7.7	50,000
1995	Kobe, Japan	6.9	5,378
2000	Indonesia	7.9	90
2001	India	7.7	>20,000

Describing Earthquake Intensity Earthquakes also can be described by the amount of damage they cause. The modified Mercalli intensity scale describes the intensity of an earthquake using the amount of structural and geologic damage in a specific location. The amount of damage done depends on the strength of the earthquake, the nature of surface material, the design of structures, and the distance from the epicenter. Under ideal conditions, only a few people would feel an intensity-I earthquake, and it would cause no damage. An intensity-IV earthquake would be felt by everyone indoors during the day but would be felt by only a few people outdoors. Pictures might fall off walls and books might fall from shelves. However, an intensity-IX earthquake would cause considerable damage to buildings and would cause cracks in the ground. An intensity-XII earthquake would cause total destruction of buildings, and objects such as cars would be thrown upward into the air. The 1994 6.8-magnitude earthquake in Northridge, California, was listed at an intensity of IX because of the damage it caused.

Liquefaction Have you ever tried to drink a thick milkshake from a cup? Sometimes the milkshake is so thick that it won't flow. How do you make the milkshake flow? You shake it. Something similar can happen to very wet soil during an earthquake. Wet soil can be strong most of the time, but the shaking from an earthquake can cause it to act more like a liquid. This is called **liquefaction.** When liquefaction occurs in soil under buildings, the buildings can sink into the soil and collapse, as shown in **Figure 17.** People living in earthquake regions should avoid building on loose soils.

Figure 17
San Francisco's Marina district suffered extensive damage from liquefaction in a 1989 earthquake because it is built on a landfilled marsh.

Tsunamis Most earthquake damage occurs when surface waves cause buildings, bridges, and roads to collapse. People living near the seashore, however, have another problem. An earthquake under the ocean causes a sudden movement of the ocean floor. The movement pushes against the water, causing a powerful wave that can travel thousands of kilometers in all directions. Far from shore, a wave caused by an earthquake is so long that a large ship might ride over it without anyone noticing. But when one of these waves breaks on a shore, as shown in **Figure 18,** it forms a towering crest that can reach 30 m in height.

Ocean waves caused by earthquakes are called seismic sea waves, or **tsunamis** (soo NAH meez). Just before a tsunami crashes onto shore, the water along a shoreline might move rapidly toward the sea, exposing a large portion of land that normally is underwater. This should be taken as a warning sign that a tsunami could strike soon, and you should head for higher ground immediately.

Because of the number of earthquakes that occur around the Pacific Ocean, the threat of tsunamis is constant. To protect lives and property, a warning system has been set up in coastal areas and for the Pacific islands to alert people if a tsunami is likely to occur. The Pacific Tsunami Warning Center, located near Hilo, Hawaii, provides warning information including predicted tsunami arrival times at coastal areas.

However, even tsunami warnings can't prevent all loss of life. In the 1960 tsunami that struck Hawaii, 61 people died when they ignored the warning to move away from coastal areas.

Figure 18
A tsunami begins over the earthquake focus. *What might happen to towns located near the shore?*

Earthquake Safety

You have learned that earthquakes can be destructive, but the damage and loss of life can be minimized. Although earthquakes cannot be predicted reliably, **Figure 19** shows where earthquakes are most likely to occur in the United States.

Knowing where large earthquakes are likely to occur helps in long-term planning. Cities in such regions can take action to prevent damage to buildings and loss of life. Many buildings withstood the 1989 San Francisco earthquake because they were built with the expectation that such an earthquake would occur someday.

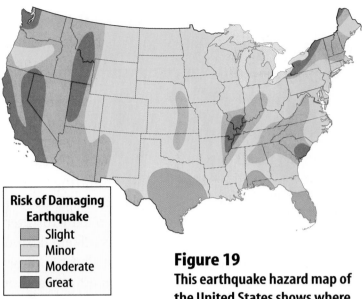

Risk of Damaging Earthquake
- Slight
- Minor
- Moderate
- Great

Figure 19
This earthquake hazard map of the United States shows where earthquakes are most likely to cause severe damage.

Math Skills Activity

Using Multiplication to Compare Earthquake Energy

Example Problem

The Richter scale is used to measure the magnitude of earthquakes. For each number increase on the Richter scale, 32 times more energy is released. How much more energy is released by a magnitude 6 earthquake than by a magnitude 3 earthquake?

Solution

1 *This is what you know:*   magnitude 6 earthquake, magnitude 3 earthquake, energy increases 32 times per magnitude number

2 *This is what you need to find out:*   amount of additional energy released

3 *This is the procedure you need to use:*   Find the difference in magnitude numbers, then use that number of <u>multiples</u> of 32 to find the amount of additional energy released.

4 *Solve the equation:*   difference in magnitude = 6 − 3 = 3
multiply 32 times itself 3 times: 32 × 32 × 32 = 32,768
32,768 times more energy is released

> **Practice Problem**
>
> Calculate the difference in the amount of energy released between a magnitude 7 earthquake and a magnitude 2 earthquake.

For more help, refer to the Math Skill Handbook.

Modeling Seismic-Safe Structures

Procedure

1. On a **tabletop,** build a structure out of **building blocks** by simply placing one block on top of another.
2. Build a second structure by wrapping sections of three blocks together with **rubber bands.** Then, wrap larger rubber bands around the entire completed structure.
3. Set the second structure on the tabletop next to the first one and pound on the side of the table with a slow, steady rhythm.

Analysis

1. Which of your two structures was better able to withstand the "earthquake" caused by pounding on the table?
2. How might the idea of wrapping the blocks with rubber bands be used in construction of supports for elevated highways?

Quake-Resistant Structures During earthquakes, buildings, bridges, and highways can be damaged or destroyed. Most loss of life during an earthquake occurs when people are trapped in or on these crumbling structures. What can be done to reduce loss of life?

Seismic-safe structures stand up to vibrations that occur during an earthquake. **Figure 20** shows how buildings can be built to resist earthquake damage. Today in California, some new buildings are supported by flexible, circular moorings placed under the buildings. The moorings are made of steel plates filled with alternating layers of rubber and steel. The rubber acts like a cushion to absorb earthquake waves. Tests have shown that buildings supported in this way should be able to withstand an earthquake measuring up to 8.3 on the Richter scale without major damage.

In older buildings, workers often install steel rods to reinforce building walls. Such measures protect buildings in areas that are likely to experience earthquakes.

✔ **Reading Check** *What are seismic-safe structures?*

Figure 20

The rubber portions of this building's moorings absorb most of the wave motion of an earthquake. The building itself only sways gently. *What purpose does the rubber serve?*

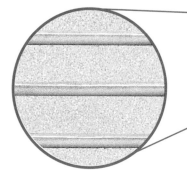

Before an Earthquake To make your home as earthquake-safe as possible, certain steps can be taken. To reduce the danger of injuries from falling objects, move heavy objects from high shelves and place them on lower shelves. Learn how to turn off the gas, water, and electricity in your home. To reduce the chance of fire from broken gas lines, make sure that hot-water heaters and other gas appliances are held securely in place, as shown in **Figure 21.** A newer method that is being used to minimize the danger of fire involves placing sensors on gas lines. The sensors automatically shut off the gas when earthquake vibrations are detected.

During an Earthquake If you're indoors, move away from windows and any objects that could fall on you. Seek shelter in a doorway or under a sturdy table or desk. If you're outdoors, stay in the open—away from power lines or anything that might fall. Stay away from buildings—chimneys or other parts of buildings could fall on you.

After an Earthquake Check water and gas lines for damage. If any are damaged, shut off the valves. If you smell gas, leave the building immediately and call authorities from a phone away from the leak area. Stay out of and away from damaged buildings. Be careful around broken glass and rubble that could contain sharp edges and wear boots or sturdy shoes to keep from cutting your feet. Finally, stay away from beaches. Tsunamis sometimes hit after the ground has stopped shaking.

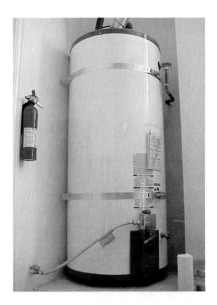

Figure 21
Securing gas water heaters to walls with sturdy straps helps reduce the danger of fires from broken gas lines.

Section 3 Assessment

1. How can you determine whether or not you live in an area where an earthquake is likely to occur?
2. What can you do to make your home more safe during an earthquake?
3. How is earthquake magnitude measured?
4. Name three ways that an earthquake can cause damage.
5. **Think Critically** How are shock absorbers on a car similar to the circular moorings used in modern earthquake-safe buildings? How do they absorb shock?

Skill Builder Activities

6. **Forming Hypotheses** Hypothesize why some earthquakes with smaller magnitudes result in more deaths than earthquakes with larger magnitudes. **For more help, refer to the** Science Skill Handbook.
7. **Solving One-Step Equations** What is the difference in energy released between an earthquake of Richter magnitude 8.5 and one of magnitude 4.5? Between one of magnitude 3.5 and one of magnitude 5.5? **For more help, refer to the** Math Skill Handbook.

Activity

Earthquake Depths

You learned earlier in this chapter that Earth's crust is broken into sections called plates. Stresses caused by movement of these plates generate energy within rocks that must be released. When this release of energy is sudden and rocks break, an earthquake occurs.

What You'll Investigate

Can a study of the foci of earthquakes tell you anything about plate movement in a particular region?

Goals
- **Observe** any connection between earthquake-focus depth and epicenter location using the data provided on the next page.
- **Describe** any observed relationship between earthquake-focus depth and the movement of plates at Earth's surface.

Materials
graph paper
pencil

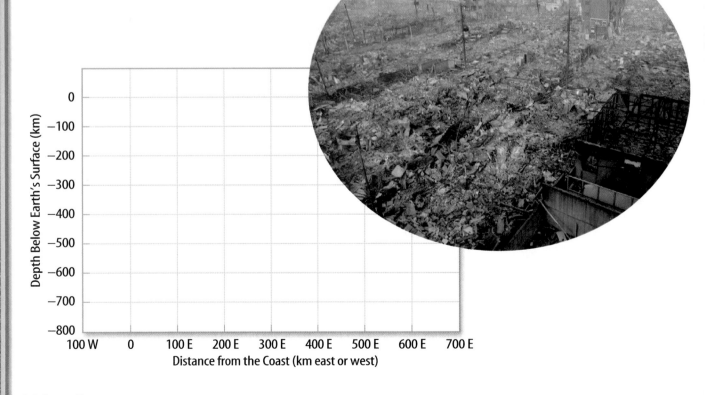

Graph axes: Depth Below Earth's Surface (km) on the vertical axis ranging from 0, −100, −200, −300, −400, −500, −600, −700, −800; Distance from the Coast (km east or west) on the horizontal axis from 100 W, 0, 100 E, 200 E, 300 E, 400 E, 500 E, 600 E, 700 E.

Procedure

1. Use graph paper and the data table on the right to make a graph plotting the depths of earthquake foci and the distances from the coast of a continent for each earthquake epicenter.

2. Use the graph on the previous page as a reference to draw your own graph. Place *Distance from the Coast* on the horizontal axis. Begin labeling at the far left with 100 km west. To the right of it should be 0 km, then 100 km east, 200 km east, 300 km east, and so on through 700 km east. What point on your graph represents the coast?

3. Label the vertical axis *Depth Below Earth's Surface.* Label the top of the graph 0 km to represent Earth's surface. Label the bottom of the vertical axis −800 km.

4. **Plot** the focus depths against the distance and direction from the coast for each earthquake in the table to the right.

Conclude and Apply

1. **Describe** any observed relationship between the location of earthquake epicenters and the depth of earthquake foci.

2. Based on the graph you have completed, hypothesize what is happening to the plates at Earth's surface in the vicinity of the plotted earthquake foci.

3. **Infer** what process is causing the earthquakes you plotted on your graph paper.

4. Hypothesize why none of the plotted earthquakes occurred below 700 km.

5. Based on what you have plotted, infer what continent these data could apply to. Explain what you based your answer on.

Focus and Epicenter Data		
Earthquake	Focus Depth (km)	Distance of Epicenter from Coast (km)
A	−55	0
B	−295	100 east
C	−390	455 east
D	−60	75 east
E	−130	255 east
F	−195	65 east
G	−695	400 east
H	−20	40 west
I	−505	695 east
J	−520	390 east
K	−385	335 east
L	−45	95 east
M	−305	495 east
N	−480	285 east
O	−665	545 east
P	−85	90 west
Q	−525	205 east
R	−85	25 west
S	−445	595 east
T	−635	665 east
U	−55	95 west
V	−70	100 west

ommunicating
Your Data

Compare your graph with those of other members of your class. **For more help, refer to the** Science Skill Handbook.

Moving Earth!

Did you know...

... The most powerful earthquake to hit the United States in recorded history shook Alaska in 1964. At 8.5 on the Richter scale, the quake shook all of Alaska for nearly 5 min, which is a long time for an earthquake. Nearly 320 km of roads near Anchorage suffered damage, and almost half of the 204 bridges had to be rebuilt.

... Snakes can sense the vibrations made by a small rodent up to 23 m away. Does this mean that they can detect vibrations prior to major earthquakes? Unusual animal behavior was observed just before a 1969 earthquake in China—an event that was successfully predicted.

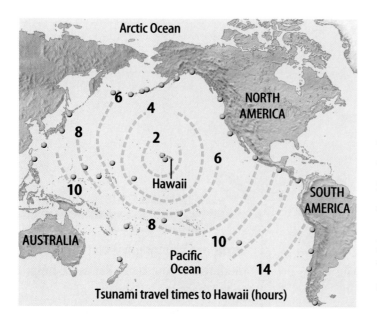

Tsunami travel times to Hawaii (hours)

... Earthquakes beneath the ocean floor can cause seismic sea waves, or tsunamis. Traveling at speeds of up to 950 km/h—as fast as a commercial jet—a tsunami can strike with little warning. Since 1945, more people have been killed by tsunamis than by the ground shaking from earthquakes.

... Tsunamis can reach heights of 30 m. A wave that tall would knock over this lighthouse.

... On December 16, 1811, a strong earthquake occurred near New Madrid, Missouri. This earthquake was so strong that it changed the course of the Mississippi River. The earthquake also was reported to have rung the bell of St. Phillip's Steeple in Charleston, South Carolina.

Do the Math

1. On the Richter scale, a whole number increase means that the height of the largest recorded seismic wave increases by ten. How much higher is the largest wave from an 8.5 earthquake than the largest wave from a 3.5 earthquake?
2. Look at the tsunami warning system map on the previous page. About how long would a tsunami triggered near the Aleutian Islands take to reach Hawaii?

Go Further

Visit the Glencoe Science Web site at **science.glencoe.com.** Research the history and effects of earthquakes in the United States. Describe how the San Francisco earthquake of 1906 stimulated earthquake research.

Reviewing Main Ideas

Section 1 Forces Inside Earth

1. Plate movements put stress on rocks. To a certain point, the rocks bend and stretch. If the force is beyond the elastic limit, the rocks might break.

2. Earthquakes are vibrations that are created when rocks break along a fault.

3. Normal faults form when rocks undergo tension. Compression produces reverse faults. Strike-slip faults result from shearing forces. *What type of fault is shown here?*

Section 2 Features of Earthquakes

1. Primary waves compress and stretch rock particles as the waves move. Secondary waves move by causing particles in rocks to move at right angles to the direction of the waves. Surface waves move in a backward rolling motion and a side-to-side swaying motion. *Which kind of earthquake wave caused the damage shown here?*

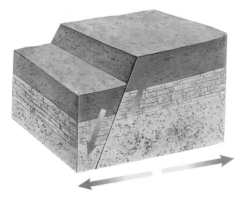

2. Scientists can locate earthquake epicenters by recording seismic waves. *Where is the epicenter of the earthquake shown here?*

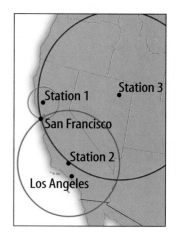

3. By observing the speeds and paths of seismic waves, scientists are able to determine the boundaries between Earth's internal layers.

Section 3 People and Earthquakes

1. A seismograph is the instrument used to measure earthquake magnitude. *What is this record of an earthquake produced by a seismograph called?*

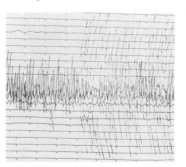

2. The magnitude of an earthquake is a measure of the energy released by the quake. The Richter scale describes how much energy an earthquake releases. The scale has no upper limit.

3. Earthquakes can cause liquefaction of wet soil and tsunamis, both of which increase the amount of structural damage produced by an earthquake.

FOLDABLES
Reading & Study Skills

After You Read

Using what you learned in this chapter, list and explain the effects of earthquakes on the inside of your Foldable.

Visualizing Main Ideas

Complete the following concept map on earthquake damage.

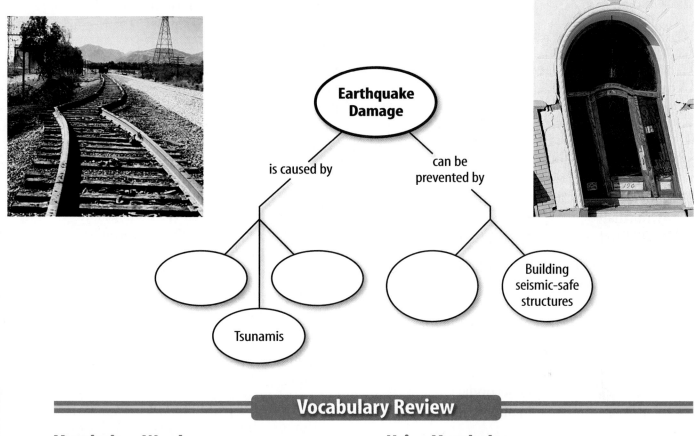

Earthquake Damage

is caused by

can be prevented by

Tsunamis

Building seismic-safe structures

Vocabulary Review

Vocabulary Words

a. earthquake
b. epicenter
c. fault
d. focus
e. liquefaction
f. magnitude
g. normal fault
h. primary wave
i. reverse fault
j. secondary wave
k. seismic wave
l. seismograph
m. strike-slip fault
n. surface wave
o. tsunami

Study Tip

Be a teacher! Gather a group of friends and assign each one a section of the chapter to teach. Teaching helps you remember and understand information.

Using Vocabulary

Replace the underlined words with the correct vocabulary words.

1. Most earthquake damage results from <u>primary waves</u>.

2. At a <u>normal fault</u>, rocks move past each other without much upward or downward movement.

3. The point on Earth's surface directly above the earthquake focus is the <u>fault</u>.

4. The measure of the energy released during an earthquake is its <u>seismograph</u>.

5. An earthquake under the ocean can cause a <u>surface wave</u> that travels thousands of kilometers.

Checking Concepts

Choose the word or phrase that best answers the question.

1. Earthquakes can occur when which of the following is passed?
   A) tension limit
   B) seismic unit
   C) elastic limit
   D) shear limit

2. When the rock above the fault surface moves down relative to the rock below the fault surface, what kind of fault forms?
   A) normal
   B) strike-slip
   C) reverse
   D) shearing

3. From which of the following do primary and secondary waves move outward?
   A) epicenter
   B) focus
   C) Moho
   D) tsunami

4. What kind of earthquake waves stretch and compress rocks?
   A) surface
   B) primary
   C) secondary
   D) shear

5. What are the slowest seismic waves?
   A) surface
   B) primary
   C) secondary
   D) pressure

6. What is the fewest number of seismograph stations that are needed to locate the epicenter of an earthquake?
   A) two
   B) three
   C) four
   D) five

7. What happens to primary waves when they pass from liquids into solids?
   A) slow down
   B) speed up
   C) stay the same
   D) stop

8. What part of a seismograph does not move during an earthquake?
   A) sheet of paper
   B) fixed frame
   C) drum
   D) pendulum

9. How much more energy does an earthquake of magnitude 7.5 release than an earthquake of magnitude 6.5?
   A) 32 times more
   B) 32 times less
   C) twice as much
   D) about half as much

10. What are the recorded lines from an earthquake called?
    A) seismograph
    B) Mercalli scale
    C) seismogram
    D) Richter scale

Thinking Critically

11. The 1960 earthquake in the Pacific Ocean off the coast of Chile caused damage and loss of life in Chile and also in Hawaii, Japan, and other areas along the Pacific Ocean border. How could this earthquake do so much damage to areas thousands of kilometers from its epicenter?

12. Why is a person who is standing outside in an open field relatively safe during a strong earthquake?

13. Explain why the pendulum of a seismograph remains at rest.

14. Tsunamis often are called tidal waves. Explain why this is incorrect.

15. Which probably would be more stable during an earthquake—a single-story wood-frame house or a brick building? Explain.

Developing Skills

16. **Communicating** Imagine you are a science reporter assigned to interview the mayor about the earthquake safety of buildings in your city. What buildings would you be most concerned about? Make a list of questions about earthquake safety that you would ask the mayor.

17. **Measuring in SI** Use an atlas and a metric ruler to answer the following question. Primary waves travel at about 6 km/s in continental crust. How long would it take a primary wave to travel from San Francisco, California, to Reno, Nevada?

18. **Interpreting Data** Use the data table below and a map of the United States to determine the location of the earthquake epicenter.

Seismograph Station Data			
Station	Latitude	Longitude	Distance from Earthquake
1	45° N	120° W	1,300 km
2	35° N	105° W	1,200 km
3	40° N	115° W	790 km

19. **Forming Hypotheses** Hypothesize how seismologists could assign magnitudes to earthquakes that occurred before modern seismographs and the Richter scale were developed.

Performance Assessment

20. **Model** Use layers of different colors of clay to illustrate the three different kinds of faults. Label each model, explaining the forces involved and the rock movement.

21. **Display** Make a display showing why data from two seismograph stations are not enough to determine the location of an earthquake epicenter.

TECHNOLOGY

Go to the Glencoe Science Web site at **science.glencoe.com** or use the **Glencoe Science CD-ROM** for additional chapter assessment.

THE PRINCETON REVIEW — Test Practice

Seismologists used the modified Mercalli intensity scale to determine the intensity of the same earthquake from four different cities. They recorded their data in the following table.

Earthquake Intensity	
City	Intensity
A	VII
B	X
C	V
D	IX

Study the table and answer the following questions.

1. According to the table, which city probably was the closest to the epicenter of the earthquake?
   A) city A
   B) city B
   C) city C
   D) city D

2. Which of the following would be an accurate conclusion based on the intensity in city B?
   F) The earthquake was not felt by very many people.
   G) The earthquake destroyed well-built wooden and stone structures.
   H) Destruction was minimal. Dishes rattled in cabinets, and pictures fell off of walls.
   J) The earthquake was only felt indoors.

Volcanoes

Every few months, villagers in Bronte, Italy, watch as Mount Etna roars to life and oozes rivers of molten lava. The village, eleven kilometers from the mountain, was out of harm's way, and no one was injured in this October 29, 1999, eruption. In this chapter, you will learn about types of volcanoes and how they form. You will learn how volcanoes affect humans and the surrounding environment and you will see the rock features they leave behind.

What do you think?

Science Journal Look at the picture below with a classmate. Discuss what you think this might be or what is happening. Here's a hint: *Not all volcanoes occur where you can see them.* Write your answer or best guess in your Science Journal.

You've seen pictures of volcanoes from the ground, but what would a volcano look like on a map? Volcanoes can be represented on maps that show the elevation of the land, as well as other important features. These maps are called topographic maps.

Map a volcano

1. Obtain half of a foam ball from your teacher and place it on the top of a table with the flat side down.

2. Using a metric ruler and a permanent marker, mark 1-cm intervals on the foam ball. Start at the base of the ball and mark up at several places around the ball.

3. Connect the marks of equal elevation by drawing a line around the ball at the 1-cm mark, at the 2-cm mark, etc.

4. Look directly down on the top of the ball and make a drawing of what you see in your Science Journal.

Observe

In your Science Journal, write a paragraph that explains how your drawing shows the general shape of a volcano. What might the lines drawn around the foam ball represent?

Before You Read

FOLDABLES
Reading & Study Skills

Making a Venn Diagram Study Fold As you prepare to read this chapter, make the following Foldable.

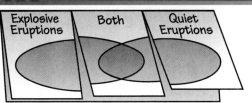

1. Place a sheet of paper in front of you so the short side is at the top. Fold the paper in half.

2. Label "Explosive Eruptions," "Quiet Eruptions," and "Both" across the front of the paper.

3. Fold both sides in to divide the paper into equal thirds. Unfold the paper so three columns show.

4. Through the top thickness of paper, cut along each of the fold lines to the top fold, forming three tabs.

5. As you read the chapter, collect information about each type of eruption under the left and right tabs. Under the middle tab, write what both types of eruptions have in common.

Volcanoes and Earth's Moving Plates

As You Read

What You'll Learn

- **Describe** how volcanoes can affect people.
- **List** conditions that cause volcanoes to form.
- **Identify** the relationship between volcanoes and Earth's moving plates.

Vocabulary

volcano   crater
vent   hot spot

Why It's Important

Volcanoes can be dangerous to people and their communities.

What are volcanoes?

A **volcano** is an opening in Earth that erupts gases, ash, and lava. Volcanic mountains form when layers of lava, ash, and other material build up around these openings. Can you name any volcanoes? Did you know that Earth has more than 600 active volcanoes?

Most Active Volcanoes Kilauea (kee low AY ah), located in Hawaii, is the world's most active volcano. For centuries, this volcano has been erupting, but not explosively. In May of 1990, most of the town of Kalapana Gardens was destroyed, but no one was hurt because the lava moved slowly and people could escape. The most recent series of eruptions from Kilauea began in January 1983 and still continues.

The island country of Iceland is also famous for its active volcanoes. It sits on an area where Earth's plates move apart and is known as the land of fire and ice. The February 26, 2000, eruption of Hekla, in Iceland, is shown in **Figure 1.**

Figure 1
This photo of the February 26, 2000 eruption of Hekla shows why Iceland is known as the land of fire and ice.

Effects of Eruptions

When volcanoes erupt, they often have direct, dramatic effects on the lives of people and their property. Lava flows destroy everything in their path. Falling volcanic ash can collapse buildings, block roads, and in some cases cause lung disease in people and animals. Sometimes, volcanic ash and debris rush down the side of the volcano. This is called a pyroclastic flow. The temperatures inside the flow can be high enough to ignite wood. When big eruptions occur, people often are forced to abandon their land and homes. People who live farther away from volcanoes are more likely to survive, but cities, towns, crops, and buildings in the area can be damaged by falling debris.

Human and Environmental Impacts The eruption of Soufrière (sew free ER) Hills volcano in Montserrat, which began in July of 1995, was one of the largest recent volcanic eruptions near North America. Geologists knew it was about to erupt, and the people who lived near it were evacuated. On June 25, 1997, large pyroclastic flows swept down the volcano. As shown in **Figure 2,** they buried cities and towns that were in their path. The eruption killed 20 people who ignored the evacuation order.

When sulfurous gases from volcanoes mix with water vapor in the atmosphere, acid rain forms. The vegetation, lakes, and streams around Soufrière Hills volcano were impacted significantly by acid rain. As the vegetation died, shown in **Figure 3,** the organisms that lived in the forest were forced to leave or also died.

Figure 3
The vegetation near the volcano on Chances Peak, on the island of Montserrat in the West Indies, was destroyed by acid rain, heat, and ash.

Astronomy
INTEGRATION

Volcanoes are not unique to Earth. Io, a moon of Jupiter, has many active volcanoes. Research to find other planets or moons that have volcanoes. Do any planets show signs of past volcanic activity?

How do volcanoes form?

What happens inside Earth to create volcanoes? Why are some areas of Earth more likely to have volcanoes than others? Deep inside Earth, heat and pressure cause rock to melt, forming liquid rock or magma. Some deep rocks already are melted. Others are hot enough that a small rise in temperature or drop in pressure can cause them to melt and form magma. What makes magma come to the surface?

Magma Forced Upward Magma is less dense than the rock around it, so it is forced slowly toward Earth's surface. You can see this process if you turn a bottle of cold syrup upside down. Watch the dense syrup force the less dense air bubbles slowly toward the top.

✔ **Reading Check**   *Why is magma forced toward Earth's surface?*

After many thousands or even millions of years, magma reaches Earth's surface and flows out through an opening called a **vent.** As lava flows out, it cools quickly and becomes solid, forming layers of igneous rock around the vent. The steep-walled depression around a volcano's vent is the **crater. Figure 4** shows magma being forced out of a volcano.

Figure 4
This cutaway diagram shows how a volcano is formed and how magma from the mantle is forced to the surface.

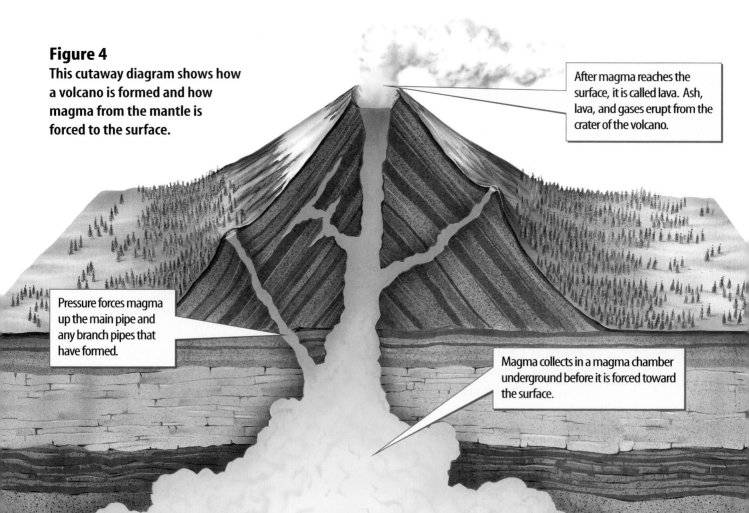

After magma reaches the surface, it is called lava. Ash, lava, and gases erupt from the crater of the volcano.

Pressure forces magma up the main pipe and any branch pipes that have formed.

Magma collects in a magma chamber underground before it is forced toward the surface.

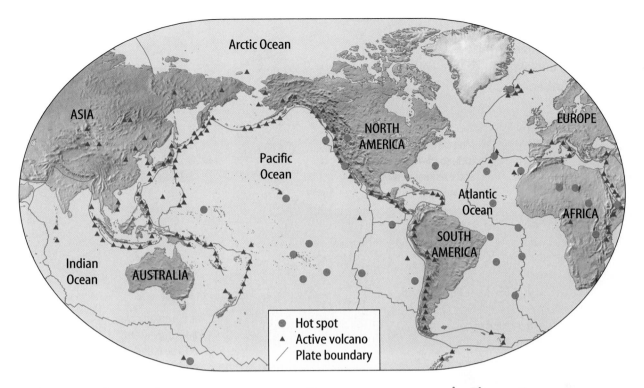

Figure 5
This map shows the locations of volcanoes, hot spots, and plate boundaries around the world. The Ring of Fire is a belt of active volcanoes that circles the Pacific Ocean.

Where do volcanoes occur?

Volcanoes often form in places where plates are moving apart, where plates are moving together, and at locations called hot spots. You can find locations of active volcanoes at plate boundaries and at hot spots on the map in **Figure 5.** Many examples can be found of volcanoes around the world that form at these three different kinds of areas. You'll explore volcanoes in Iceland, on the island of Montserrat, and in Hawaii.

Divergent Plate Boundaries Iceland is a large island in the North Atlantic Ocean. It is near the Arctic Circle and therefore has some glaciers. Iceland has volcanic activity because it sits on top of the Mid-Atlantic Ridge.

The Mid-Atlantic Ridge is a divergent plate boundary, which is an area where Earth's plates are moving apart. When plates separate, they form long, deep cracks called rifts. Lava flows from these rifts and is cooled quickly by seawater. **Figure 6** shows how magma rises at rifts to form new volcanic rock. As more lava flows and hardens, it builds up on the seafloor. Sometimes, the volcanoes and rift eruptions rise above sea level, forming islands such as Iceland. In 1963, the new island Surtsey was formed during a volcanic eruption.

Figure 6
This diagram shows how volcanic activity occurs where Earth's plates move apart.

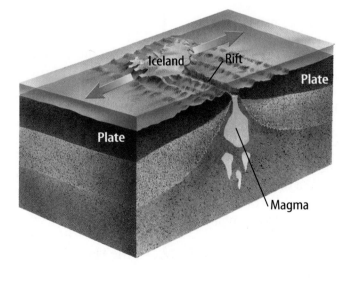

Figure 7
Volcanoes can form where plates collide and one plate slides below the other.

Magma is less dense than rock, so it is forced upward and eventually erupts from the volcano.

As the oceanic plate slides downward, rock melts and forms magma.

Convergent Plate Boundaries Places where Earth's plates move together are called convergent plate boundaries. They include areas where an oceanic plate slides below a continental plate as in **Figure 7,** and where one oceanic plate slides below another oceanic plate. The Andes in South America began forming when an oceanic plate started sliding below a continental plate. Volcanoes that form on convergent plate boundaries tend to erupt more violently than other volcanoes do.

Magma forms when the plate sliding below another plate and the overlying mantle melt partially. The magma then is forced upward to the surface, forming volcanoes like Soufrière Hills on the island of Montserrat.

Hot Spots The Hawaiian Islands are forming as a result of volcanic activity. However, unlike Iceland, they haven't formed at a plate boundary. The Hawaiian Islands are in the middle of the Pacific Plate, far from its edges. What process could be forming them?

It is thought that some areas at the boundary between Earth's mantle and core are unusually hot. Hot rock at these areas is forced toward the crust where it melts partially to form a **hot spot.** The Hawaiian Islands sit on top of a hot spot under the Pacific Plate. Magma has broken through the crust to form several volcanoes. The volcanoes that rise above the water form the Hawaiian Islands, shown in **Figure 8A.**

Figure 8
The Hawaiian Islands are actually volcanoes.

A This satellite photo shows the Hawaiian Islands. *Why are they in a relatively straight line?*

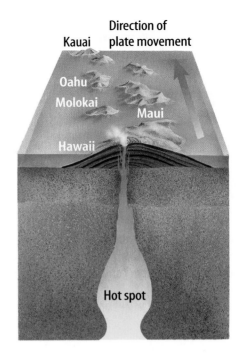

Kauai

Oahu
Molokai

Maui

Hawaii

Direction of plate movement

Hot spot

B This illustration shows how the Hawaiian Islands were formed over a hot spot.

The Hawaiian Islands As you can see in **Figure 8,** the Hawaiian Islands are all in a line. This is because the Pacific Plate is moving over a stationary hot spot. Kauai, the oldest Hawaiian island, was once located where the big island, Hawaii, is situated today. As the plate moved, Kauai moved away from the hot spot and became dormant. As the Pacific Plate continued to move, the islands of Oahu, Molokai, Maui, and Hawaii were formed. The Hawaiian Islands formed over a period of about 5 million years.

Section 1 Assessment

1. How are volcanoes related to Earth's moving plates?

2. Hot lava flows are not the only danger associated with active volcanoes. What effects can pyroclastic flows have on people?

3. Why does lava cool rapidly along a mid-ocean ridge?

4. Describe processes that are occurring to cause Soufrière Hills volcano to erupt.

5. **Think Critically** If the Pacific Plate stopped moving, what might happen to the island of Hawaii?

Skill Builder Activities

6. **Concept Mapping** Make a concept map that shows how the Hawaiian Islands formed. Use the following phrases: *volcano forms, plate moves, volcano becomes dormant,* and *new volcano forms.* **For more help, refer to the** Science Skill Handbook.

7. **Communicating** Scientists were able to predict approximately when Mount Pinatubo in the Philippines would erupt in 1991. In your Science Journal, write a report on equipment used to predict volcanic eruptions. **For more help, refer to the** Science Skill Handbook.

2 Types of Volcanoes

As You Read

What You'll Learn

- **Explain** how the explosiveness of a volcanic eruption is related to the silica and water vapor content of its magma.
- **List** three forms of volcanoes.

Vocabulary

shield volcano   cinder cone volcano
tephra   composite volcano

Why It's Important

If you know the type of volcano, you can predict how it will erupt.

Figure 9
A calm day in Washington state was shattered suddenly when Mount St. Helens erupted on May 18, 1980, as shown in this sequence of photographs.

What controls eruptions?

Some volcanic eruptions are explosive, like those from Soufrière Hills volcano, Mount Pinatubo, and Mount St. Helens. In others, the lava quietly flows from a vent, as in the Kilauea eruptions. What causes these differences?

Two important factors control whether an eruption will be explosive or quiet. One factor is the amount of water vapor and other gases that are trapped in the magma. The second factor is how much silica is present in the magma. Silica is a compound composed of the elements silicon and oxygen.

Trapped Gases When you shake a soft-drink container and then quickly open it, the pressure from the gas in the drink is released suddenly, spraying the drink all over. In the same way, gases such as water vapor and carbon dioxide are trapped in magma by the pressure of the surrounding magma and rock. As magma nears the surface, it is under less pressure. This allows the gas to escape from the magma. Gas escapes easily from some magma during quiet eruptions. However, gas that builds up to high pressures eventually causes explosive eruptions such as the one shown in **Figure 9.**

 A 8:32 A.M.

B 38 seconds later

Water Vapor The magma at some convergent plate boundaries contains a lot of water vapor. This is because oceanic plate material and some of its water slide under other plate material at some convergent plate boundaries. The trapped water vapor in the magma can cause explosive eruptions.

Composition of Magma

The second major factor that affects the nature of the eruption is the composition of the magma. Magma can be divided into two major types—silica poor and silica rich.

Quiet Eruptions Magma that is relatively low in silica is called basaltic magma. It is fluid and produces quiet, non-explosive eruptions such as those at Kilauea. This type of lava pours from volcanic vents and runs down the sides of a volcano. As this *pahoehoe* (pa-HOY-hoy) lava cools, it forms a ropelike structure. If the same lava flows at a lower temperature, a stiff, slowly moving *aa* (AH-ah) lava forms. In fact, you can walk right up to some aa lava flows on Kilauea.

Figure 10 shows some different types of lava. These quiet eruptions form volcanoes over hot spots such as the Hawaiian volcanoes. Basaltic magmas also flow from rift zones, which are long, deep cracks in Earth's surface. Many lava flows in Iceland are of this type. Because basaltic magma is fluid when it is forced upward in a vent, trapped gases can escape easily in a nonexplosive manner, sometimes forming lava fountains. Lavas that flow underwater form pillow lava formations. They consist of rock structures shaped like tubes, balloons, or pillows.

SCIENCE Online

Research Visit the Glencoe Science Web site at **science.glencoe.com** to learn more about Kilauea volcano in Hawaii. Draw a map of Hawaii that shows the location of Kilauea.

C 42 seconds later

D 53 seconds later

Figure 10

Lava rarely travels faster than a few kilometers an hour. Therefore, it poses little danger to people. However, homes and property can be damaged. On land, there are two main types of lava flows—aa (AH ah) and pahoehoe (pa HOY hoy). When lava comes out of cracks in the ocean floor, it is called pillow lava. The lava cooling here came from a volcanic eruption on the island of Hawaii.

Aa flows, like this one on Mount Etna in Italy, carry sharp angular chunks of rock called scoria. Aa flows move slowly and are intensely hot.

Pillow lava occurs where lava oozes out of cracks in the ocean floor. It forms pillow-shaped lumps as it cools. Pillow lava is the most common type of lava on Earth.

Pahoehoe flows, like this one near Kilauea's Mauna Ulu Crater in Hawaii, are more fluid than aa flows. They develop a smooth skin and form rope-

Figure 11

A Magmas that are rich in silica produce violent eruptions, such as this one in Alaska.
B This color enhanced view of volcanic ash, from a 10 million year old volcano in Nebraska, shows the glass particles that make up ash.

Magnification: 450×

Explosive Magma Silica-rich, or granitic, magma on the other hand produces explosive eruptions such as those at Soufrière Hills volcano. This magma sometimes forms where Earth's plates are moving together and one plate slides under another. As the plate that is sliding under the other goes deeper, some rock is melted. The magma is forced upward by denser surrounding rock, comes in contact with the crust, and becomes enriched in silica. Silica-rich granitic magma is thick, and gas gets trapped inside, causing pressure to build up. When an explosive eruption occurs, as shown in **Figure 11,** the gases expand rapidly, often carrying pieces of lava in the explosion.

Reading Check *What type of magmas produce violent eruptions?*

Some magmas have an andesitic composition. Andesitic magma is more silica rich than basaltic magma is, but it is less silica rich than granitic magma. It often forms at convergent plate boundaries where one plate slides under the other. Because of their higher silica content, they also erupt more violently than basaltic magmas. One of the biggest eruptions in recorded history, Krakatau, was primarily andesitic in composition. The word *andesitic* comes from the Andes, which are mountains located along the western edge of South America, where andesite rock is common. Many of the volcanoes encircling the Pacific Ocean also are made of andesite.

Health
INTEGRATION

When volcanoes erupt, ash often is spread over a great distance. People who live near volcanoes must be careful not to inhale too much of the ash particles because the particles can cause respiratory problems. In your Science Journal, describe what people can do to prevent exposure to volcanic ash.

Forms of Volcanoes

A volcano's form depends on whether it is the result of a quiet or an explosive eruption and the type of lava it is made of—basaltic, granitic, or andesitic (intermediate). The three basic types of volcanoes are shield volcanoes, cinder cone volcanoes, and composite volcanoes.

Shield Volcano Quiet eruptions of basaltic lava spread out in flat layers. The buildup of these layers forms a broad volcano with gently sloping sides called a **shield volcano,** as seen in **Figure 12.** The Hawaiian Islands are examples of shield volcanoes. Basaltic lava also can flow onto Earth's surface through large cracks called fissures. This type of eruption forms flood basalts, not volcanoes, and accounts for the greatest volume of erupted volcanic material. The basaltic lava flows over Earth's surface, covering large areas with thick deposits of basaltic igneous rock when it cools. The Columbia Plateau located in the northwestern United States was formed in this way. Much of the new seafloor that originates at mid-ocean ridges forms as underwater flood basalts.

Cinder Cone Volcano Explosive eruptions throw lava and rock high into the air. Bits of rock or solidified lava dropped from the air are called **tephra** (TEH fruh). Tephra varies in size from volcanic ash, to cinders, to larger rocks called bombs and blocks. When tephra falls to the ground, it forms a steep-sided, loosely packed **cinder cone volcano,** as seen in **Figure 13.**

Figure 12
A shield volcano like Mauna Loa, shown here, is formed when lava flows from one or more vents without erupting violently.

Vent

Magma

Paricutín On February 20, 1943, a Mexican farmer learned about cinder cones when he went to his cornfield. He noticed that a hole in his cornfield that had been there for as long as he could remember was giving off smoke. Throughout the night, hot glowing cinders were thrown high into the air. In just a few days, a cinder cone several hundred meters high covered his cornfield. This is the volcano named Paricutín.

Composite Volcano Some volcanic eruptions can vary between quiet and violent, depending on the amount of trapped gases and how rich in silica the magma is. An explosive period can release gas and ash, forming a tephra layer. Then, the eruption can switch to a quieter period, erupting lava over the top of the tephra layer. When this cycle of lava and tephra is repeated over and over in alternating layers, a **composite volcano** is formed. Composite volcanoes, shown in **Figure 14,** are found mostly where Earth's plates come together and one plate slides below the other. Soufrière Hills volcano is an example. As you can see in **Table 1** on the next page, many things affect eruptions and the form of a volcano.

Figure 13
Paricutín is a large, cinder cone volcano located in Mexico.

Figure 14
Mount Rainier in the state of Washington is an example of a composite volcano.

Violent Eruptions Soufrière Hills volcano formed as ocean floor of the North American Plate and the South American Plate slid beneath the Caribbean Plate, causing magma to form. Successive eruptions of lava and tephra produced the majestic composite volcanoes that tower above the surrounding landscape on Montserrat and other islands in the Lesser Antilles. Before the 1995 eruption, silica-rich magma rose and was trapped beneath the surface. As the magma was forced toward Earth's surface, the pressure on the underlying magma was released. This started a series of eruptions that were still continuing in the year 2001.

Table 1 Thirteen Selected Eruptions

Volcano and Location	Year	Type	Eruptive Force	Magma Content		Ability of Magma to Flow	Products of Eruption
				Silica	H$_2$O		
Mount Etna, Sicily	1669	composite	moderate	high	low	medium	lava, ash
Tambora, Indonesia	1815	cinder cone	high	high	high	low	cinders, ash
Krakatau, Indonesia	1883	composite	high	high	high	low	cinders, ash
Mount Pelée, Martinique	1902	cinder cone	high	high	high	low	gas, ash
Vesuvius, Italy	1906	composite	moderate	high	low	medium	lava, ash
Mount Katmai, Alaska	1912	composite	high	high	high	low	lava, ash, gas
Paricutìn, Mexico	1943	cinder cone	moderate	high	low	medium	ash, cinders
Surtsey, Iceland	1963	shield	moderate	low	low	high	lava, ash
Mount St. Helens, Washington	1980	composite	high	high	high	low	gas, ash
Kilauea, Hawaii	1983	shield	low	low	low	high	lava
Mount Pinatubo, Philippines	1991	composite	high	high	high	low	gas, ash
Soufrière Hills, Montserrat	1995	composite	high	high	high	low	gas, ash, rocks
Popocatépetl, Mexico	2000	composite	moderate	high	low	medium	gas, ash

Figure 15
Not much was left after Krakatau erupted in 1883.

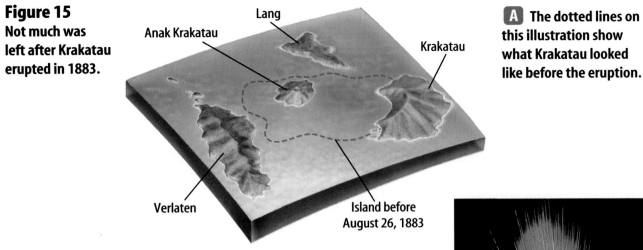

A The dotted lines on this illustration show what Krakatau looked like before the eruption.

Krakatau One of the most violent eruptions in recent times occurred on an island in the Sunda Straits near Indonesia in August of 1883. Krakatau, a volcano on the island, erupted with such force that the island disappeared as shown in **Figure 15A.** Most of the island collapsed into the emptied magma chamber. The noise of the eruption was so loud that it woke people in Australia and was heard as far away as 4,653 km from the island. Ash from the eruption fell in Singapore, which is 840 km to the north, and the area around the volcano was in complete darkness for 24 h. More than 36,000 people were killed, most by the giant tsunami waves created by the eruption. Global temperatures were lowered as much as 1.2°C by particles blown into the atmosphere and didn't return to normal until 1888.

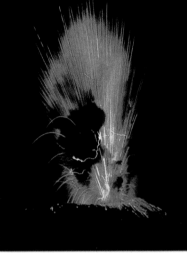

B Anak Krakatau formed in the early 1900s. The name means "Child of Krakatau."

Section 2 Assessment

1. Some eruptions are quiet and others are violent. What causes this difference?

2. Compare and contrast the different types of lava.

3. How is a composite volcano like a shield and a cinder cone volcano?

4. Describe how the Hawaiian Islands formed in the Pacific Ocean.

5. **Think Critically** In 1883, Krakatau in Indonesia erupted. Infer which kind of lava Krakatau erupted—lava rich in silica or lava low in silica. Support your inference using data in **Table 1.**

Skill Builder Activities

6. **Comparing and Contrasting** Use **Table 1** to compare and contrast Kilauea in Hawaii and Mount Pinatubo in the Philippines. **For more help, refer to the** Science Skill Handbook.

7. **Calculating Ratios** When Mount St. Helens erupted in 1980, about 1.3 km³ of material were ejected from the volcano. Tambora in Indonesia gave off 131 km³ of material in 1815. How many times larger was the volume of material given off by Tambora? Based on your information, which volcano did the most damage? **For more help, refer to the** Math Skill Handbook.

Activity

Identifying Types of Volcanoes

You have learned that certain properties of magma are related to the type of eruption and the form of the volcano that will develop. Do this activity to see how to make and use a table that relates the properties of magma to the form of volcano that develops.

What You'll Investigate
Are the silica and water content of a magma related to the form of volcano that develops?

Materials
Table 1 of thirteen selected eruptions
paper
pencil

Goals
- ■ **Determine** any relationship between the ability of magma to flow and eruptive force.
- ■ **Determine** any relationship between magma composition and eruptive force.

Procedure
1. Copy the graph shown above.
2. Using the information from **Table 1,** plot the magma content for each of the volcanoes listed by writing the name of the basic type of volcano in the correct spot on the graph.
3. After you plot all 13 volcanoes, analyze the patterns of volcanic types on the diagram to answer the questions.

Conclude and Apply
1. What relationship appears to exist between the ability of the magma to flow and the eruptive force of the volcano?

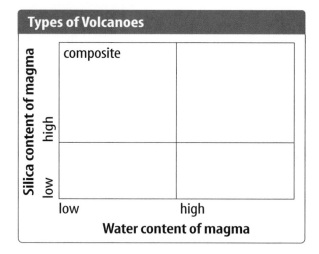

Types of Volcanoes

2. Which would be more liquidlike: magma that flows easily or magma that flows with difficulty?
3. What relationship appears to exist between the silica or water content of the magma and the nature of the material ejected from the volcano?
4. How is the ability of a magma to flow related to its silica content?
5. **Infer** which of the two variables, silica or water content, appears to have the greater effect on the eruptive force of the volcano.
6. **Describe** the relationship that appears to exist between the silica and water content of the magma and the type of volcano that is produced.

*C*ommunicating Your Data

Create a poster that shows the relationship between magma composition and the type of volcano formed. **For more help, refer to the** Science Skill Handbook.

3 Igneous Rock Features

Intrusive Features

You can observe volcanic eruptions because they occur at Earth's surface. However, far more activity occurs underground. In fact, most magma never reaches Earth's surface to form volcanoes or to flow as flood basalts. This magma cools slowly underground and produces underground rock bodies that could become exposed later at Earth's surface by erosion. These rock bodies are called intrusive igneous rock features. There are several different types of intrusive features. Some of the most common are batholiths, sills, dikes, and volcanic necks. What do intrusive igneous rock bodies look like? You can see illustrations of these features in **Figure 16.**

As You Read

What You'll Learn

- **Describe** intrusive igneous rock features and how they form.
- **Explain** how a volcanic neck and a caldera form.

Vocabulary

batholith   volcanic neck
dike   caldera
sill

Why It's Important

Many features formed underground by igneous activity are exposed at Earth's surface by erosion.

Figure 16
This diagram shows intrusive and other features associated with volcanic activity. *Which features shown are formed above ground? Which are formed by intrusive activities?*

Volcanic neck

Lava flow from fissure

Composite volcano

Batholith

Sill

Magma chamber

Dike

Batholiths The largest intrusive igneous rock bodies are **batholiths.** They can be many hundreds of kilometers in width and length and several kilometers thick. Batholiths form when magma bodies that are being forced upward from inside Earth cool slowly and solidify before reaching the surface. However, not all of them remain hidden inside Earth. Some batholiths have been exposed at Earth's surface by many years of erosion. The granite domes of Yosemite National Park are the remains of a huge batholith that stretches across much of the length of California.

Math Skills Activity

Classifying Igneous Rocks

Igneous rocks are classified into three types depending on the amount of silica they contain. Basaltic rocks contain approximately 45 percent to 52 percent silica. Andesitic, or intermediate, rocks contain about 52 percent to 66 percent silica, and granitic rocks have more than 66 percent silica. The lighter the color is, the higher the silica content is.

Example Problem

A 900-kg block of igneous rock contains 630 kg of silica. Calculate the percent of silica in the rock to classify it.

Solution

1 *This is what you know:*   rock = 900 kg
                                    silica = 630 kg

2 *This is what you need to find:*   The percentage of silica: x

3 *This is the equation you need to use:*   Mass of silica / mass of rock = x / 100

4 *Solve the equation for x:* $x = (630 \text{ kg}/900 \text{ kg}) \times 100$
                                     $x = 70$ percent, therefore, the rock is granitic.

Check your answer by dividing it by 100, then multiplying by 900. Did you get the given amount of silica?

Practice Problems

1. A 250-kg boulder of basalt contains 125 kg of silica. Use the classification system to determine whether basalt is light or dark.
2. Andesite is an intermediate, medium-colored rock with a silica content ranging from 52 percent to 66 percent. About how many kilograms of silica would you predict to be in a 68-kg boulder of andesite?

For help with solving equations, refer to the Math Skill Handbook.

Dikes and Sills Magma sometimes squeezes into cracks in rock below the surface. This is like squeezing toothpaste into the spaces between your teeth. Magma that is forced into a crack that cuts across rock layers and hardens is called a **dike.** Magma that is forced into a crack parallel to rock layers and hardens is called a **sill.** These features are shown in **Figures 17A** and **17B.** Most dikes and sills run from a few meters to hundreds of meters long.

Other Features

When a volcano stops erupting, the magma hardens inside the vent. Erosion, usually by water and wind, begins to wear away the volcano. The cone is much softer than the solid igneous rock in the vent. Thus, the cone erodes first, leaving behind the solid igneous core as a **volcanic neck.** Devil's Tower in Wyoming, shown in **Figure 17C,** might be an example of a volcanic neck.

SCIENCE *Online*

Research Visit the Glencoe Science Web site at **science.glencoe.com** to learn more about igneous rock features. Share your research with your class.

A A sill is formed when magma is forced between rock layers.

Figure 17
Igneous features can form in many different sizes and shapes.

B The vertical dikes shown here near Shiprock, New Mexico, were formed when magma squeezed into vertical cracks in the surrounding rock layers.

C Some people have suggested that Devil's Tower is a volcanic neck.

Calderas Sometimes after an eruption, the top of a volcano can collapse, as seen in **Figure 18.** This produces a large depression called a **caldera.** Crater Lake in Oregon, shown in **Figure 19,** is a caldera that filled with water and is now a lake. Crater Lake formed after the violent eruption and destruction of Mount Mazama about 7,000 years ago.

Figure 18
Calderas are formed when the top of a volcano collapses.

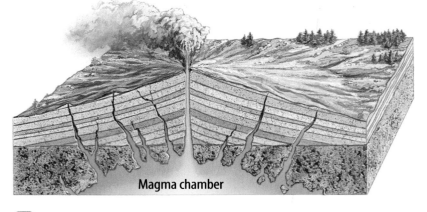

A Magma rises, causing volcanic activity to occur.

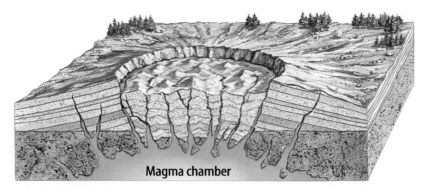

B The magma chamber partially empties, causing rock to collapse into the emptied chamber below the surface. This forms a circular-shaped caldera.

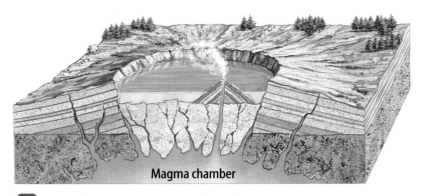

C Crater Lake in Oregon formed when water collected in the circular space left when surface material collapsed.

Figure 19
Wizard Island in Crater Lake is a cinder cone volcano that erupted after the formation of the caldera. *What causes a caldera to form?*

Igneous Features Exposed You have learned in this chapter that Earth's surface is built up and worn down continually. The surface of Earth is built up by volcanoes. Also, igneous rock is formed when magma hardens below ground. Eventually, the processes of weathering and erosion wear down rock at the surface, exposing features like batholiths, dikes, and sills.

 Reading Check *What exposes igneous features that formed below the surface?*

Section 3 Assessment

1. What's the difference between a caldera and a crater?

2. Describe how a sill forms. How is it different from a dike?

3. What is a volcanic neck and how does it form?

4. Explain how a batholith forms.

5. **Think Critically** Why are the large, granite dome features of Yosemite National Park in California considered to be intrusive volcanic features when they are exposed at the surface?

Skill Builder Activities

6. **Comparing and Contrasting** Compare and contrast dikes, sills, batholiths, and volcanic necks. **For more help, refer to the** Science Skill Handbook.

7. **Using Graphics Software** Use the graphics software available on your computer to produce an illustration of igneous rock features based on **Figure 16.** Be sure to include intrusive features and features that form above ground. **For more help, refer to the** Technology Skill Handbook.

Activity
Design Your Own Experiment

How do calderas form?

A caldera is a depression formed when the top of a volcano collapses after an eruption. What might cause the top of a volcano to collapse? What would happen if the magma inside the magma chamber suddenly were removed?

Recognize the Problem

How does the removal of magma from the magma chamber affect a volcano?

Form a Hypothesis

Based on your reading about volcanoes, state a hypothesis about what would happen if the magma inside the magma chamber of a volcano were suddenly removed.

Goals
- **Design** a volcano setup that will demonstrate how a caldera could form.
- **Observe** what happens during trials with your volcano setup.
- **Describe** what you observe.

Possible Materials
small box
small balloon
paper
newspaper
flour
plastic tubing
clamp for tubing
tape
scissors

Safety Precautions

Test Your Hypothesis

Plan

1. As a group, agree upon the hypothesis and identify which results will support the hypothesis.
2. **Design** a volcano that allows you to test your hypothesis. What materials will you use to build your volcano?
3. What will you remove from inside your volcano to represent the loss of magma? How will you remove it?
4. Where will you place your volcano? What will you do to minimize messes?
5. **Identify** all constants, variables, and controls of the experiment.

Do

1. Make sure your teacher approves your plan before you start.
2. **Construct** your volcano with any features that will be required to test your hypothesis.
3. **Conduct** one or more appropriate trials to test your hypothesis. Record any observations that you make and any other data that are appropriate to test your hypothesis.

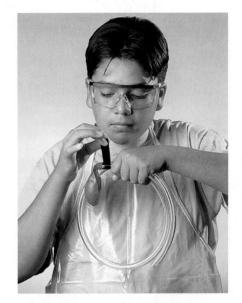

Analyze Your Data

1. **Describe** in words or with a drawing what your volcano looked like before you began.
2. What happened to your volcano during the experiment that you conducted? Did its appearance change?
3. **Describe** in words or with a drawing what your volcano looked like after the trial.
4. What other observations did you make?
5. **Describe** any other data that you recorded.

Draw Conclusions

1. Did your observations support your hypothesis? Explain.
2. How was your demonstration similar to what might happen to a real volcano? How was it different?

Communicating Your Data

Make a poster with diagrams and descriptions of how a caldera forms. Use your visual aid to **describe** caldera formation to students in another class.

Buried in Ash

A long-forgotten city is accidentally found after 2,000 years

In the heat of the Italian Sun, a tired farmer wipes his brow. The farmer has spent the morning digging a new well for water. The hole is deep and the ground is dusty. Heaving a sigh, the farmer thrusts the shovel into the ground one more time.

But instead of hitting water, the shovel strikes something hard. It is a slab of smooth white marble.

This richly decorated public bath was unearthed at Herculaneum (large photo).

The small photo shows excavated ruins with Mount Vesuvius in the background.

The farmer didn't know it at the time (the early 1700s), but that marble was the first clue that something very big and very important lay beneath the farm fields. Under the ground people walked on every day, lay the ancient city of Herculaneum (her kew LAY nee um). The city, and its neighbor Pompeii (pom PAY) had been buried for more than 1,600 years. Why? Because on another summer day, August 24, 79 A.D., to be exact, Mount Vesuvius, a nearby volcano, erupted and buried both cities with pumice, rocks, mud, and ash.

An archaeologist excavates a skeleton in Herculaneum.

Back in Time

The sun shone over the town of Herculaneum on that August morning almost 2,000 years ago. Nestled at the foot of the mountain, overlooking the Gulf of Naples, it was a peaceful place. But at about 1 P.M., that peace was shattered forever.

With massive force, the peak of Vesuvius exploded, sending six cubic kilometers of ash and pumice into the sky. Hours later, a fiery surge made its way from the volcano to the city. These pyroclastic flows continued as gray pumice fell from the sky. Buildings were crushed and buried by falling ash and pumice. Within six hours, much of Herculaneum was totally buried under the flows. After six surges from Vesuvius, the deadly eruption ceased. But the city had disappeared under approximately 21 m of ash, rock, and mud.

A City Vanishes

More than 3,600 people were killed in the natural disaster. Scientists believe that most were killed by the pyroclastic surges. Many died trying to protect their faces from the air that was filled with hot ash. Those lucky enough to escape returned to find no trace of their city. Over hundreds of years, grass and fields covered Herculaneum, erasing it from human memory. Eventually, a town called Resina was built on the site.

In the last couple of hundred years, archaeologists have unearthed colorful and perfectly preserved mosaics and an amazing library with ancient scrolls in excellent condition. Archaeologists found skeletons and voids that were filled with plaster to form casts of people who died when Vesuvius erupted. Visitors to the site can see a Roman woman, a teen-aged girl, and a soldier with his sword still in his hand.

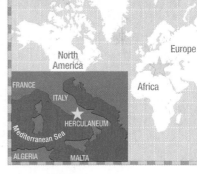

Much of Herculaneum still lies buried beneath thick layers of volcanic ash, and archaeologists still are digging to expose more of the ruins. Their work is helping scientists better understand everyday life in an ancient Italian town. But if it weren't for a farmer's search for water, Herculaneum might not have been discovered at all!

Reviewing Main Ideas

Section 1 Volcanoes and Earth's Moving Plates

1. Volcanoes can be dangerous to people because they can cause deaths and destroy property.

2. Rocks in the crust and mantle melt to form magma, which is forced toward Earth's surface. When the magma flows through vents, it's called lava and forms volcanoes. *What is happening to this lava?*

3. Volcanoes can form over hot spots when magma flows onto Earth's surface. Sometimes the lava builds up from the seafloor to form an island. Volcanoes also form when Earth's plates pull apart or come together.

Section 2 Types of Volcanoes

1. The three types of volcanoes are shield volcanoes, cinder cone volcanoes, and composite volcanoes. *Which type of volcano is pictured below?*

2. Shield volcanoes produce quiet eruptions. Cinder cone and composite volcanoes can produce explosive eruptions.

3. Some lavas are thin and flow easily, producing quiet eruptions. Other lavas are thick and stiff, producing violent eruptions.

4. Water vapor and silica in magma add to its explosiveness.

Section 3 Igneous Rock Features

1. Intrusive igneous rock bodies such as batholiths, dikes, and sills form when magma solidifies underground.

2. Batholiths are the most massive igneous rock bodies. Dikes form when magma squeezes into cracks, cutting across rock layers. Sills form when magma squeezes in between rock layers. *Which rock feature is shown in this photo?*

3. A caldera forms when the top of a volcano collapses, forming a large depression. Crater Lake is a caldera in Oregon.

After You Read

FOLDABLES
Reading & Study Skills

Use your Foldable to help review the similarities and differences between quiet and explosive volcanic eruptions. Decide what type of volcano produces each style of eruption.

Visualizing Main Ideas

Complete the following concept map on types of volcanic eruptions.

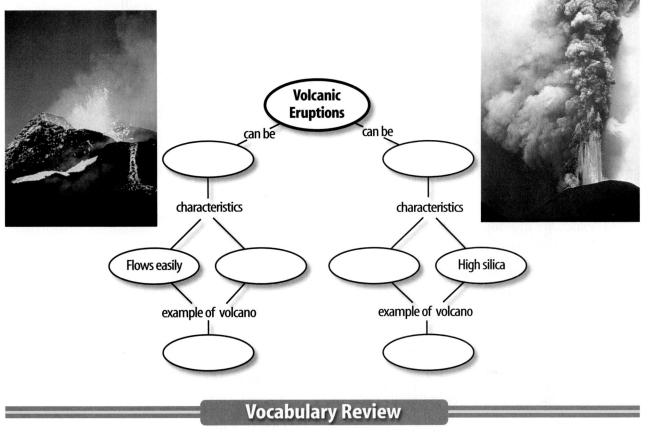

Vocabulary Review

Vocabulary Words

a. batholith
b. caldera
c. cinder cone volcano
d. composite volcano
e. crater
f. dike
g. hot spot
h. shield volcano
i. sill
j. tephra
k. vent
l. volcanic neck
m. volcano

THE PRINCETON REVIEW **Study Tip**

When you encounter new vocabulary, write it down in a sentence. This will help you understand, remember, and use new vocabulary words.

Using Vocabulary

Each of the following sentences is false. Make the sentence true by replacing each underlined word(s) with the correct vocabulary word(s).

1. A broad volcano with gently sloping sides is called a <u>composite volcano</u>.

2. <u>Sills</u> are bits of rock or solidified lava dropped from the air after a volcanic eruption.

3. Magma squeezed into a horizontal crack between rock layers is called a <u>caldera</u>.

4. The steep-walled depression around a volcano's vent is called a <u>tephra</u>.

5. Magma squeezed into a vertical crack across rock layers is called a <u>crater</u>.

Chapter 6 Assessment

Checking Concepts

Choose the word or phrase that best answers the question.

1. What type of boundary is associated with composite volcanoes?
   A) plates moving apart
   B) plates sticking and slipping
   C) plates moving together
   D) plates sliding past each other

2. Why is Hawaii made of volcanoes?
   A) Plates are moving apart.
   B) A hot spot exists.
   C) Plates are moving together.
   D) Rift zones exist.

3. What kind of magmas produce violent volcanic eruptions?
   A) those rich in silica
   B) those that are fluid
   C) those forming shield volcanoes
   D) those rich in iron

4. Magma that is low in silica generally produces what kind of eruptions?
   A) thick
   C) quiet
   B) caldera
   D) explosive

5. What is made entirely of tephra?
   A) shield volcano
   C) cinder cone volcano
   B) caldera
   D) composite volcano

6. What kind of volcano is Kilauea?
   A) shield volcano
   C) cinder cone volcano
   B) composite
   D) caldera cone volcano

7. What is magma that hardens in a crack cutting across rock layers called?
   A) sill
   C) volcanic neck
   B) dike
   D) batholith

8. What is the largest intrusive igneous rock body?
   A) dike
   C) sill
   B) volcanic neck
   D) batholith

9. Which describes bits of material that fall to Earth after an eruption?
   A) dike
   C) tephra
   B) sand
   D) sill

10. What is the process that formed Soufrière Hills volcano on Montserrat?
    A) plates sticking and slipping
    B) caldera formation
    C) plates sliding sideways
    D) plates moving together

Thinking Critically

11. Explain how glaciers and volcanoes can exist on Iceland.

12. What kind of eruption is produced when basaltic lava that is low in silica flows from a volcano? Explain.

13. How are volcanoes related to earthquakes?

14. Misti is a volcano in Peru. Peru is on the western edge of South America. How might this volcano have formed?

15. Describe the layers of a composite volcano. Which layers represent violent eruptions?

Developing Skills

16. **Classifying** Classify Fuji, which has steep sides and is made of layers of silica-rich lava and ash.

17. **Measuring in SI** The base of the volcano Mauna Loa is about 5,000 m below sea level. The total height of the volcano is 9,170 m. What percentage of the volcano is above sea level? Below sea level?

18. **Comparing and Contrasting** Compare and contrast shield volcanoes, cinder cone volcanoes, and composite volcanoes.

19. Interpreting Scientific Illustrations Look at the map below. The Hawaiian Islands and Emperor Seamounts were formed when the Pacific Plate moved over a fixed hot spot. If the Emperor chain trends in a direction different from the Hawaiian Islands, what can you infer about the Pacific Plate?

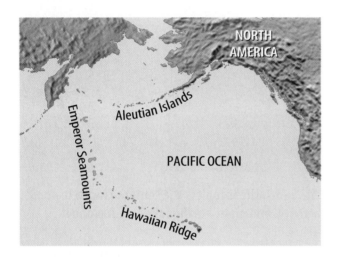

20. Concept Mapping Make a network tree concept map about where volcanoes can occur. Include the following words and phrases: *hot spots, divergent plate boundaries, convergent plate boundaries, volcanoes, can occur, examples, Iceland, Soufrière Hills,* and *Hawaiian Islands.*

Performance Assessment

21. Poster Make a poster of the three basic types of volcanoes. Label them and indicate what type of eruption occurs from each one.

TECHNOLOGY

Go to the Glencoe Science Web site at **science.glencoe.com** or use the **Glencoe Science CD-ROM** for additional chapter assessment.

THE PRINCETON REVIEW — Test Practice

A scientist who studies volcanoes brought the following information to a lecture he did at a middle school for the entire sixth grade.

Study the table and answer the following questions.

Data for Various Volcanoes		
Volcano	**Silica Content in Magma**	**Trapped Gases in Magma**
1	Medium to High	High
2	Low	Medium
3	Low to Medium	Medium
4	High	High
5	Low	Medium

1. According to this information, which volcano probably had the most violent eruption?
A) Volcano 1
B) Volcano 2
C) Volcano 4
D) Volcano 5

2. Based on this information, choose the most reasonable hypothesis.
F) The higher the silica content is, the stronger the eruption is.
G) The lower the trapped gas content is, the stronger the eruption is.
H) The lower the silica content is, the stronger the eruption is.
J) The higher the trapped gas content is, the weaker the eruption is.

Read the passage carefully. Then read the questions that follow the passage. Decide which is the best answer to each question.

Test-Taking Tip Identify whether this passage is fictional or informational.

Earthquakes and Volcanoes

Earthquakes are destructive and potentially fatal natural disasters. Geologists have been working to learn what they can about earthquakes in order to better protect property and to save human lives.

Scientists know that many earthquakes occur because tectonic plates interact with one another at plate boundaries. They also know that many earthquakes occur every day all over the world. They forecast that major earthquakes will occur sometime in the future in certain regions, such as along the San Andreas Fault near Parkfield, California. But scientists cannot predict exactly when and where an earthquake will occur.

One approach to forecasting earthquakes is known as paleoseismology. Paleoseismology involves the study of past movement of rock and sediment along faults. Motion along a fault results in an earthquake, therefore, studying this movement is one way to study ancient earthquake occurrences.

This movement can be measured in the field by observing shifted rock and sediment along a fault. If this displaced sediment can be dated, the time at which the earthquake occurred also can be estimated. With information on several past earthquakes, scientists can estimate how long, on average, time intervals are between the earthquakes. This is one way to estimate how many earthquakes might occur along a fault over a period of time.

Although scientists are not yet able to predict earthquakes, the information gained from their research advances the field.

The San Andreas fault is the center of much tectonic activity, and many earthquakes occur along it.

1. Based on the information in this passage, what can the reader conclude?
   A) Earthquakes always occur during heavy rainstorms.
   B) Earthquakes can not be predicted reliably, but they can be forecasted over estimated periods of time.
   C) Earthquakes are extremely rare natural occurrences.
   D) Earthquakes do not affect property and human lives very much.

2. What does the information in this passage suggest?
   F) Studies of displacement along faults can provide information for forecasting earthquakes.
   G) Studying faults allows scientists to determine the time and date of the next earthquake.
   H) Paleoseismology is the study of fossils along the San Andreas fault.
   J) Scientists do not know what causes earthquakes.

Reasoning and Skills

Read each question and decide which is the best answer.

Igneous Rocks		
Formed	Light-colored	Dark-colored
Below Earth's Surface	granite	gabbro
At Earth's Surface	rhyolite	basalt

1. According to the chart, lava that flows onto the surface from a volcano should cool to form the dark-colored rock _____.
  A) granite   **C)** rhyolite
  B) gabbro   **D)** basalt

Test-Taking Tip   Reread the question and think about the color of the rock and where the rock was formed.

2. Earth's crust is estimated to be composed of about 46% oxygen, 28% silicon, 8% aluminum, and 18% other elements. Which area of the graph represents aluminum?
  F) Q
  G) R
  H) S
  J) T

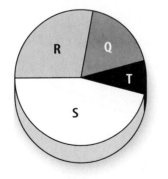

Test-Taking Tip   Think about the quantities of the element the question refers to compared to the sizes of each portion of the graph.

3. All of these statements are true about volcanoes EXCEPT _____.
  A) Three basic types of volcanoes are shield, cinder cone, and composite.
  B) The amount of water vapor, other gases, and silica in magma determine what kind of eruption will take place.
  C) The lower the silica content in magma, the more explosive an eruption will be.
  D) A volcano's form depends upon the type of eruption and the composition of erupted material.

Test-Taking Tip   Think about the different kinds of volcanoes, how they form, and what factors influence the type of eruption.

Consider this question carefully before writing your answer on a separate sheet of paper.

4. Volcanic eruptions transform the environment around them. Discuss some of the ways in which volcanoes change Earth's surface.

Test-Taking Tip   Consider the details of specific eruptions that you have learned about.

Student Resources

Student Resources

Building Stones

Since ancient times, people have used naturally available materials such as stone and wood to construct their homes, places of worship, palaces, and other buildings and monuments. As early as 2500 B.C., the Egyptians had learned how to cut and transport large blocks of limestone from mountainsides for use in the construction of the pyramids. The Maya also used stone to construct their magnificent cities. Since those times, stone has remained a popular building material not only for its beauty but also for its durability and strength.

Types of Building Stone and Some Famous Stone Monuments

The choice of stone used in construction depends on the purpose, availability, cost, and properties of the particular stone. On the following pages are some of the most common building stones, one or more of which was used in the construction of many of the famous buildings in the world. By using this Field Guide, you can learn which stones were used to build these structures. You also can learn to identify the stones used to construct other monuments, buildings, and structures in your neighborhood or town.

Granite

Granite's mineral composition determines its color. For example, some granites are pink because they contain the mineral potassium feldspar. The minerals in granite resist wear and tear caused by wind and precipitation.

The pink color of granite in Enchanted Rock is caused by the presence of microcline, a variety of potassium feldspar.

Marischal College

Built in the late 1890s in Aberdeen, Scotland, Marischal College is one of the largest granite buildings in the world. Constructed in a gothic style, the building has an ornate façade and represents unusually intricate sculpture work for granite.

Field Activity

Take a walk around your town and find a public building or structure made of stone. Then use this field guide to identify the stones used to construct that building. Record your findings in your Science Journal. Visit the Glencoe Science Web site at **science.glencoe.com** to find out which stones were used to build other famous buildings.

Limestone

Limestone commonly ranges in color from white to gray, but other colors are possible. Compared to other stones, limestone is not resistant when exposed to air and water in humid climates. However, it still is used as a building stone, especially where it is available locally.

Indiana limestone

Marble

In its purest form marble is white, but impurities produce other colors. The impurities often weave throughout marble, producing a beautiful mosaic appearance. Many marbles are composed mainly of calcite or dolomite. The softness of these minerals allows marble to be carved easily.

Italian marble

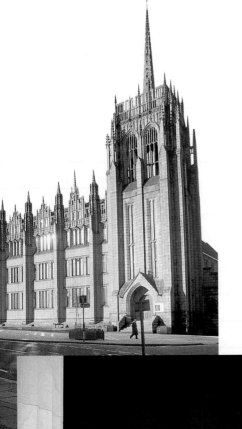

Lincoln Memorial

The Lincoln Memorial in Washington, D.C., was built to honor Abraham Lincoln. The exterior of the building, which is constructed in the Greek style, is made of marble quarried from Colorado and Tennessee. The walls and columns inside are made of Indiana limestone, and the ceiling, floor, and platform are made of marble from Tennessee and Alabama. Lincoln's statue was carved using white marble from Georgia.

Sandstone

The sand grains in sandstone give it a coarse, rustic appearance. Although sandstone often is white or tan, mineral cements can color the rock vivid shades of orange or red. When sand grains are well-cemented and composed mainly of resistant minerals such as quartz, sandstones can be excellent building materials.

Uluru (OO lew rew), the world's largest outcropping of sandstone, was previously called Ayers Rock.

Piece of sandstone

The Red Fort

Built by the Shah Jahan in the seventeenth century, the Red Fort in Delhi, India, is made of red sandstone. The octagonal, or eight-sided, fort contains palaces, gardens, army barracks, and other buildings, including a stable for the king's horses and elephants. The entire fort is surrounded by a large wall.

The Gloddfa Ganol Slate Mine in Wales is the world's largest slate mine.

Slate

The color of slate varies, but it is usually gray, black, green, or red. The color is determined by the stones' organic and mineral content. Slate contains fine-grained, well-compacted clay minerals and mica which make it water-tight and easy to separate into layers. These properties make slate ideal to use for roofing and paving.

Mud Island Park

The flat and compact nature of slate make it a natural choice for this River Walk in Mud Island Park, Memphis, Tennessee. A street map of Memphis was etched into the surface of the slate.

Mud Island Park River Walk

Organizing Information

As you study science, you will make many observations and conduct investigations and experiments. You will also research information that is available from many sources. These activities will involve organizing and recording data. The quality of the data you collect and the way you organize it will determine how well others can understand and use it. In **Figure 1,** the student is obtaining and recording information using a thermometer.

Putting your observations in writing is an important way of communicating to others the information you have found and the results of your investigations and experiments.

Researching Information

Scientists work to build on and add to human knowledge of the world. Before moving in a new direction, it is important to gather the information that already is known about a subject. You will look for such information in various reference sources. Follow these steps to research information on a scientific subject:

Step 1 Determine exactly what you need to know about the subject. For instance, you might want to find out what happened when Mount St. Helens erupted in 1980.

Step 2 Make a list of questions, such as: When did the eruption begin? How long did it last? What kind of material was expelled and how much?

Step 3 Use multiple sources such as textbooks, encyclopedias, government documents, professional journals, science magazines, and the Internet.

Step 4 List where you found the sources. Make sure the sources you use are reliable and the most current available.

Figure 1
Collecting data is one way to gather information directly.

Evaluating Print and Nonprint Sources

Not all sources of information are reliable. Evaluate the sources you use for information, and use only those you know to be dependable. For example, suppose you live in an area where earthquakes are common and you want to know what to do to keep safe. You might find two Web sites on earthquake safety. One Web site contains "Earthquake Tips" written by a company that sells metal scrapings to help secure your hot-water tank to the wall. The other is a Web page on "Earthquake Safety" written by the U.S. Geological Survey. You would choose the second Web site as the more reliable source of information.

In science, information can change rapidly. Always consult the most current sources. A 1985 source about the Moon would not reflect the most recent research and findings.

Skill Handbooks

Interpreting Scientific Illustrations

As you research a science topic, you will see drawings, diagrams, and photographs. Illustrations help you understand what you read. Some illustrations are included to help you understand an idea that you can't see easily by yourself. For instance, you can't see the layers of Earth, but you can look at a diagram of Earth's layers, as labeled in **Figure 2,** that helps you understand what the layers are and where they are located. Visualizing a drawing helps many people remember details more easily. Illustrations also provide examples that clarify difficult concepts or give additional information about the topic you are studying.

Most illustrations have a label or caption. A label or caption identifies the illustration or provides additional information to better explain it. Can you find the caption or labels in **Figure 2?**

Figure 2
This cross section shows a slice through Earth's interior and the positions of its layers.

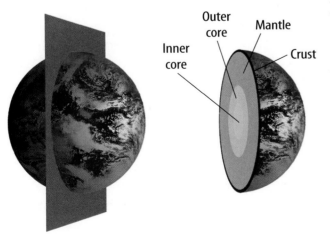

Venn Diagram

Although it is not a concept map, a Venn diagram illustrates how two subjects compare and contrast. In other words, you can see the characteristics that the subjects have in common and those that they do not.

The Venn diagram in **Figure 3** shows the relationship between two types of rocks made from the same basic chemical. Both rocks share the chemical calcium carbonate. However, due to the way they are formed, one rock is the sedimentary rock limestone, and the other is the metamorphic rock marble.

Concept Mapping

If you were taking a car trip, you might take some sort of road map. By using a map, you begin to learn where you are in relation to other places on the map.

A concept map is similar to a road map, but a concept map shows relationships among ideas (or concepts) rather than places. It is a diagram that visually shows how concepts are related. Because a concept map shows relationships among ideas, it can make the meanings of ideas and terms clear and help you understand what you are studying.

Overall, concept maps are useful for breaking large concepts down into smaller parts, making learning easier.

Figure 3
A Venn diagram shows how objects or concepts are alike and how they are different.

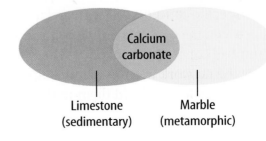

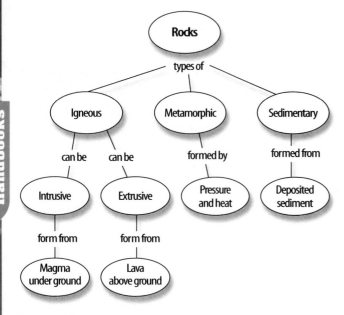

Figure 4
A network tree shows how concepts or objects are related.

Network Tree Look at the network tree in **Figure 4,** that shows the three main types of rock. A network tree is a type of concept map. Notice how some words are in ovals while others are written across connecting lines. The words inside the ovals are science terms or concepts. The words written on the connecting lines describe the relationships between the concepts.

When constructing a network tree, write the topic on a note card or piece of paper. Write the major concepts related to that topic on separate note cards or pieces of paper. Then arrange them in order from general to specific. Branch the related concepts from the major concept and describe the relationships on the connecting lines. Continue branching to more specific concepts. Write the relationships between the concepts on the connecting lines until all concepts are mapped. Then examine the network tree for relationships that cross branches, and add them to the network tree.

Events Chain An events chain is another type of concept map. It models the order of items or their sequence. In science, an events chain can be used to describe a sequence of events, the steps in a procedure, or the stages of a process.

When making an events chain, first find the one event that starts the chain. This event is called the *initiating event.* Then, find the next event in the chain and continue until you reach an outcome. Suppose you are asked to describe why and how a sound might make an echo. You might draw an events chain such as the one in **Figure 5.** Notice that connecting words are not necessary in an events chain.

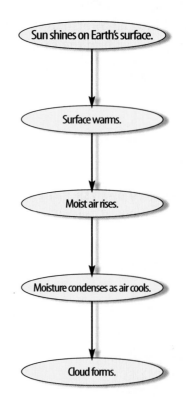

Figure 5
Events chains show the order of steps in a process or event.

Cycle Map

Cycle Map A cycle concept map is a specific type of events chain map. In a cycle concept map, the series of events does not produce a final outcome. Instead, the last event in the chain relates back to the beginning event.

You first decide what event will be used as the beginning event. Once that is decided, you list events in order that occur after it. Words are written between events that describe what happens from one event to the next. The last event in a cycle concept map relates back to the beginning event. The number of events in a cycle concept varies but is usually three or more. Look at the cycle map shown in **Figure 6.**

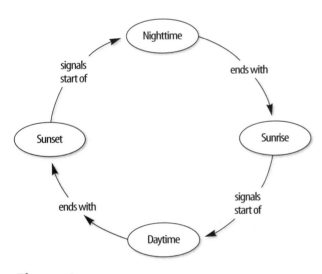

Figure 6
A cycle map shows events that occur in a cycle.

Spider Map A type of concept map that you can use for brainstorming is the spider map. When you have a central idea, you might find you have a jumble of ideas that relate to it but are not necessarily clearly related to each other. The spider map on mining in **Figure 7** shows that if you write these ideas outside the main concept, then you can begin to separate and group unrelated terms so they become more useful.

Figure 7
A spider map allows you to list ideas that relate to a central topic but not necessarily to one another.

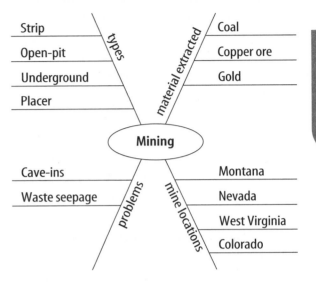

Writing a Paper

You will write papers often when researching science topics or reporting the results of investigations or experiments. Scientists frequently write papers to share their data and conclusions with other scientists and the public. When writing a paper, use these steps.

Step 1 Assemble your data by using graphs, tables, or a concept map. Create an outline.

Step 2 Start with an introduction that contains a clear statement of purpose and what you intend to discuss or prove.

Step 3 Organize the body into paragraphs. Each paragraph should start with a topic sentence, and the remaining sentences in that paragraph should support your point.

Step 4 Position data to help support your points.

Step 5 Summarize the main points and finish with a conclusion statement.

Step 6 Use tables, graphs, charts, and illustrations whenever possible.

You might say the work of a scientist is to solve problems. When you decide to find out why one corner of your yard is always soggy, you are problem solving, too. You might observe that the corner is lower than the surrounding area and has less vegetation growing in it. You might decide to see whether planting some grass will keep the corner drier.

Scientists use orderly approaches to solve problems. The methods scientists use include identifying a question, making observations, forming a hypothesis, testing a hypothesis, analyzing results, and drawing conclusions.

Scientific investigations involve careful observation under controlled conditions. Such observation of an object or a process can suggest new and interesting questions about it. These questions sometimes lead to the formation of a hypothesis. Scientific investigations are designed to test a hypothesis.

Identifying a Question

The first step in a scientific investigation or experiment is to identify a question to be answered or a problem to be solved. You might be interested in knowing why a rock like the one in **Figure 8** looks the way it does.

Figure 8
When you find a rock, you might ask yourself, "How did this rock form?"

Forming Hypotheses

Hypotheses are based on observations that have been made. A hypothesis is a possible explanation based on previous knowledge and observations.

Perhaps a scientist has observed that thunderstorms happen more often on hot days than on cooler days. Based on these observations, the scientist can make a statement that he or she can test. The statement is a hypothesis. The hypothesis could be: *Warm temperatures cause thunderstorms*. A hypothesis has to be something you can test by using an investigation. A testable hypothesis is a valid hypothesis.

Predicting

When you apply a hypothesis, or general explanation, to a specific situation, you predict something about that situation. First, you must identify which hypothesis fits the situation you are considering. People use predictions to make everyday decisions. Based on previous observations and experiences, you might form a prediction that if warm temperatures cause thunderstorms, then more thunderstorms will occur in summer months than in spring months. Someone could use these predictions to plan when to take a camping trip or when to schedule an outdoor activity.

Testing a Hypothesis

To test a hypothesis, you need a procedure. A procedure is the plan you follow in your experiment. A procedure tells you what materials to use, as well as how and in what order to use them. When you follow a procedure, data are generated that support or do not support the original hypothesis statement.

For example, premium gasoline costs more than regular gasoline. Does premium gasoline increase the efficiency or fuel mileage of your family car? You decide to test the hypothesis: "If premium gasoline is more efficient, then it should increase the fuel mileage of my family's car." Then you write the procedure shown in **Figure 9** for your experiment and generate the data presented in the table below.

Figure 9
A procedure tells you what to do step by step.

> **Procedure**
> 1. Use regular gasoline for two weeks.
> 2. Record the number of kilometers between fill-ups and the amount of gasoline used.
> 3. Switch to premium gasoline for two weeks.
> 4. Record the number of kilometers between fill-ups and the amount of gasoline used.

Gasoline Data			
Type of Gasoline	Kilometers Traveled	Liters Used	Liters per Kilometer
Regular	762	45.34	0.059
Premium	661	42.30	0.064

These data show that premium gasoline is less efficient than regular gasoline in one particular car. It took more gasoline to travel 1 km (0.064) using premium gasoline than it did to travel 1 km using regular gasoline (0.059). This conclusion does not support the hypothesis.

Are all investigations alike? Keep in mind as you perform investigations in science that a hypothesis can be tested in many ways. Not every investigation makes use of all the ways that are described on these pages, and not all hypotheses are tested by investigations. Scientists encounter many variations in the methods that are used when they perform experiments. The skills in this handbook are here for you to use and practice.

Identifying and Manipulating Variables and Controls

In any experiment, it is important to keep everything the same except for the item you are testing. The one factor you change is called the independent variable. The factor that changes as a result of the independent variable is called the dependent variable. Always make sure you have only one independent variable. If you allow more than one, you will not know what causes the changes you observe in the dependent variable. Many experiments also have controls—individual instances or experimental subjects for which the independent variable is not changed. You can then compare the test results to the control results.

For example, in the fuel-mileage experiment, you made everything the same except the type of gasoline that was used. The driver, the type of automobile, and the type of driving were the same throughout. In this way, you could be sure that any mileage differences were caused by the type of fuel—the independent variable. The fuel mileage was the dependent variable.

If you could repeat the experiment using several automobiles of the same type on a standard driving track with the same driver, you could make one automobile a control by using regular gasoline over the four-week period.

Collecting Data

Whether you are carrying out an investigation or a short observational experiment, you will collect data, or information. Scientists collect data accurately as numbers and descriptions and organize it in specific ways.

Observing Scientists observe items and events, then record what they see. When they use only words to describe an observation, it is called qualitative data. For example, a scientist might describe the color, texture, or odor of a substance produced in a chemical reaction. Scientists' observations also can describe how much there is of something. These observations use numbers, as well as words, in the description and are called quantitative data. For example, if a sample of the element gold is described as being "shiny and very dense," the data are clearly qualitative. Quantitative data on this sample of gold might include "a mass of 30 g and a density of 19.3 g/cm^3." Quantitative data often are organized into tables. Then, from information in the table, a graph can be drawn. Graphs can reveal relationships that exist in experimental data.

When you make observations in science, you should examine the entire object or situation first, then look carefully for details. If you're looking at a rock sample, for instance, check the general color and pattern of the rock before using a hand lens to examine the small mineral grains that make up its underlying structure. Remember to record accurately everything you see.

Scientists try to make careful and accurate observations. When possible, they use instruments such as microscopes, metric rulers, graduated cylinders, thermometers, and balances. Measurements provide numerical data that can be repeated and checked.

Sampling When working with large numbers of objects or a large population, scientists usually cannot observe or study every one of them. Instead, they use a sample or a portion of the total number. To *sample* is to take a small, representative portion of the objects or organisms of a population for research. By making careful observations or manipulating variables within a portion of a group, information is discovered and conclusions are drawn that might apply to the whole population.

Estimating Scientific work also involves estimating. To *estimate* is to make a judgment about the amount or the number of something without measuring every part of an object or counting every member of a population. Scientists first measure or count the amount or number in a small sample. A chemist, for example, might remove a 10-g piece of a large rock that is rich in copper ore. Then the chemist can determine the percentage of copper by mass and multiply that percentage by the mass of the rock to get an estimate of the total mass of copper in the rock. See **Figure 10** for another example.

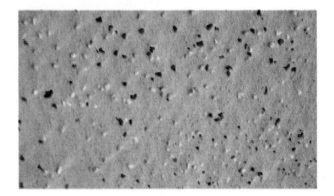

Figure 10
In a 1 m² frame positioned on a beach, count all the pebbles that you can see on the surface that are longer than 2.5 cm. Multiply this number by the area of the beach. This will give you an estimate for the total number of pebbles on the beach.

Measuring in SI

The metric system of measurement was developed in 1795. A modern form of the metric system, called the International System, or SI, was adopted in 1960. SI provides standard measurements that all scientists around the world can understand.

The metric system is convenient because unit sizes vary by multiples of 10. When changing from smaller units to larger units, divide by a multiple of 10. When changing from larger units to smaller, multiply by a multiple of 10. To convert millimeters to centimeters, divide the millimeters by 10. To convert 30 mm to centimeters, divide 30 by 10 (30 mm equal 3 cm).

Prefixes are used to name units. Look at the table below for some common metric prefixes and their meanings. Do you see how the prefix *kilo-* attached to the unit *gram* is *kilogram*, or 1,000 g?

Metric Prefixes			
Prefix	**Symbol**	**Meaning**	
kilo-	k	1,000	thousand
hecto-	h	100	hundred
deka-	da	10	ten
deci-	d	0.1	tenth
centi-	c	0.01	hundredth
milli-	m	0.001	thousandth

Now look at the metric ruler shown in **Figure 11.** The centimeter lines are the long, numbered lines, and the shorter lines are millimeter lines.

When using a metric ruler, line up the 0-cm mark with the end of the object being measured, and read the number of the unit where the object ends. In this instance, it would be 4.5 cm.

Figure 11
This metric ruler shows centimeters and millimeter divisions.

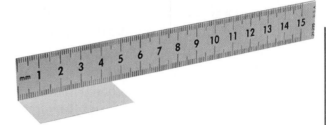

Liquid Volume In some science activities, you will measure liquids. The unit that is used to measure liquids is the liter. A liter has the volume of 1,000 cm³. The prefix *milli-* means "thousandth (0.001)." A milliliter is one thousandth of 1 L, and 1 L has the volume of 1,000 mL. One milliliter of liquid completely fills a cube measuring 1 cm on each side. Therefore, 1 mL equals 1 cm³.

You will use beakers and graduated cylinders to measure liquid volume. A graduated cylinder, as illustrated in **Figure 12,** is marked from bottom to top in milliliters. This one contains 79 mL of a liquid.

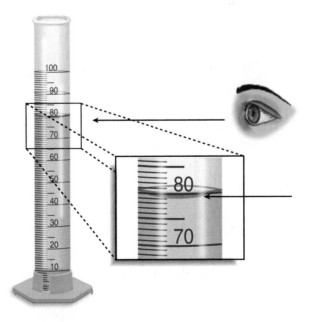

Figure 12
Graduated cylinders measure liquid volume.

Mass Scientists measure mass in grams. You might use a beam balance similar to the one shown in **Figure 13.** The balance has a pan on one side and a set of beams on the other side. Each beam has a rider that slides on the beam.

Before you find the mass of an object, slide all the riders back to the zero point. Check the pointer on the right to make sure it swings an equal distance above and below the zero point. If the swing is unequal, find and turn the adjusting screw until you have an equal swing.

Place an object on the pan. Slide the largest rider along its beam until the pointer drops below zero. Then move it back one notch. Repeat the process on each beam until the pointer swings an equal distance above and below the zero point. Sum the masses on each beam to find the mass of the object. Move all riders back to zero when finished.

Figure 13
A triple beam balance is used to determine the mass of an object.

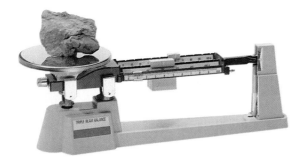

You should never place a hot object on the pan or pour chemicals directly onto the pan. Instead, find the mass of a clean container. Remove the container from the pan, then place the chemicals in the container. Find the mass of the container with the chemicals in it. To find the mass of the chemicals, subtract the mass of the empty container from the mass of the filled container.

Making and Using Tables

Browse through your textbook and you will see tables in the text and in the activities. In a table, data, or information, are arranged so that they are easier to understand. Activity tables help organize the data you collect during an activity so results can be interpreted.

Making Tables To make a table, list the items to be compared in the first column and the characteristics to be compared in the first row. The title should clearly indicate the content of the table, and the column or row heads should tell the reader what information is found in there. The table below lists materials collected for recycling on three weekly pick-up days. The inclusion of kilograms in parentheses also identifies for the reader that the figures are mass units.

Recyclable Materials Collected During Week			
Day of Week	Paper (kg)	Aluminum (kg)	Glass (kg)
Monday	5.0	4.0	12.0
Wednesday	4.0	1.0	10.0
Friday	2.5	2.0	10.0

Using Tables How much paper, in kilograms, is being recycled on Wednesday? Locate the column labeled "Paper (kg)" and the row "Wednesday." The information in the box where the column and row intersect is the answer. Did you answer "4.0"? How much aluminum, in kilograms, is being recycled on Friday? If you answered "2.0," you understand how to read the table. How much glass is collected for recycling each week? Locate the column labeled "Glass (kg)" and add the figures for all three rows. If you answered "32.0," then you know how to locate and use the data provided in the table.

Recording Data

To be useful, the data you collect must be recorded carefully. Accuracy is key. A well-thought-out experiment includes a way to record procedures, observations, and results accurately. Data tables are one way to organize and record results. Set up the tables you will need ahead of time so you can record the data right away.

Record information properly and neatly. Never put unidentified data on scraps of paper. Instead, data should be written in a notebook like the one in **Figure 14.** Write in pencil so information isn't lost if your data gets wet. At each point in the experiment, record your data and label it. That way, your information will be accurate and you will not have to determine what the figures mean when you look at your notes later.

Figure 14
Record data neatly and clearly so it is easy to understand.

Recording Observations

It is important to record observations accurately and completely. That is why you always should record observations in your notes immediately as you make them. It is easy to miss details or make mistakes when recording results from memory. Do not include your personal thoughts when you record your data. Record only what you observe to eliminate bias. For example, when you record the time required for five students to climb the same set of stairs, you would note which student took the longest time. However, you would not refer to that student's time as "the worst time of all the students in the group."

Making Models

You can organize the observations and other data you collect and record in many ways. Making models is one way to help you better understand the parts of a structure you have been observing or the way a process for which you have been taking various measurements works.

Models often show things that are too large or too small for normal viewing. For example, you normally won't see the inside of an atom. However, you can understand the structure of the atom better by making a three-dimensional model of an atom. The relative sizes, the positions, and the movements of protons, neutrons, and electrons can be explained in words. An atomic model made of a plastic-ball nucleus and pipe-cleaner electron shells can help you visualize how the parts of the atom relate to each other.

Other models can be devised on a computer. Some models, such as those that illustrate the chemical combinations of different elements, are mathematical and are represented by equations.

Making and Using Graphs

After scientists organize data in tables, they might display the data in a graph that shows the relationship of one variable to another. A graph makes interpretation and analysis of data easier. Three types of graphs are the line graph, the bar graph, and the circle graph.

Line Graphs A line graph like in **Figure 15** is used to show the relationship between two variables. The variables being compared go on two axes of the graph. For data from an experiment, the independent variable always goes on the horizontal axis, called the *x*-axis. The dependent variable always goes on the vertical axis, called the *y*-axis. After drawing your axes, label each with a scale. Next, plot the data points.

A data point is the intersection of the recorded value of the dependent variable for each tested value of the independent variable. After all the points are plotted, connect them.

Bar Graphs Bar graphs compare data that do not change continuously. Vertical bars show the relationships among data.

To make a bar graph, set up the *y*-axis as you did for the line graph. Draw vertical bars of equal size from the *x*-axis up to the point on the *y*-axis that represents the value of *x*.

Figure 16

The amount of aluminum collected for recycling during one week can be shown as a bar graph or circle graph.

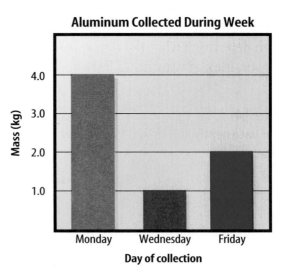

Circle Graphs A circle graph uses a circle divided into sections to display data as parts (fractions or percentages) of a whole. The size of each section corresponds to the fraction or percentage of the data that the section represents. So, the entire circle represents 100 percent, one-half represents 50 percent, one-fifth represents 20 percent, and so on.

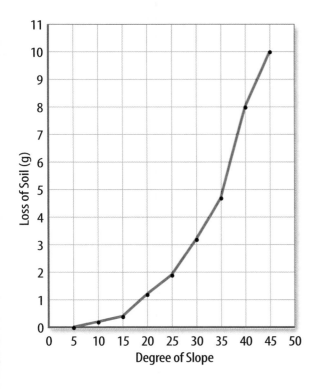

Figure 15

This line graph shows the relationship between degree of slope and the loss of soil in grams from a container during an experiment.

Analyzing Results

To determine the meaning of your observations and investigation results, you will need to look for patterns in the data. You can organize your information in several of the ways that are discussed in this handbook. Then you must think critically to determine what the data mean. Scientists use several approaches when they analyze the data they have collected and recorded. Each approach is useful for identifying specific patterns in the data.

Forming Operational Definitions

An operational definition defines an object by showing how it functions, works, or behaves. Such definitions are written in terms of how an object works or how it can be used; that is, they describe its job or purpose.

For example, a ruler can be defined as a tool that measures the length of an object (how it can be used). A ruler also can be defined as something that contains a series of marks that can be used as a standard when measuring (how it works).

Classifying

Classifying is the process of sorting objects or events into groups based on common features. When classifying, first observe the objects or events to be classified. Then select one feature that is shared by some members in the group but not by all. Place those members that share that feature into a subgroup. You can classify members into smaller and smaller subgroups based on characteristics.

How might you classify a group of rocks? You might first classify them by color, putting all of the black, white, and red rocks into separate groups. Within each group, you could then look for another common feature to classify further, such as size or whether the rocks have sharp or smooth edges.

Remember that when you classify, you are grouping objects or events for a purpose. For example, classifying rocks can be the first step in identifying them. You might know that obsidian is a black, shiny rock with sharp edges. To find it in a large group of rocks, you might start with the classification scheme mentioned. You'll locate obsidian within the group of black, sharp-edged rocks that you separate from the rest. Pumice could be located by its white color and by the fact that it contains many small holes called vesicles. Keep your purpose in mind as you select the features to form groups and subgroups.

Figure 17
Color is one of many characteristics that are used to classify rocks.

Comparing and Contrasting

Observations can be analyzed by noting the similarities and differences between two or more objects or events that you observe. When you look at objects or events to see how they are similar, you are comparing them. Contrasting is looking for differences in objects or events. The table below compares and contrasts the characteristics of two minerals.

Mineral Characteristics		
Mineral	Graphite	Gold
Color	black	bright yellow
Hardness	1–2	2.5–3
Luster	metallic	metallic
Uses	pencil "lead"	jewelry, electronics

Recognizing Cause and Effect

Have you ever heard a loud pop right before the power went out and then suggested that an electric transformer probably blew out? If so, you have observed an effect and inferred a cause. The event is the effect, and the reason for the event is the cause.

When scientists are unsure of the cause of a certain event, they design controlled experiments to determine what caused it.

Interpreting Data

The word *interpret* means "to explain the meaning of something." Look at the problem originally being explored in an experiment and figure out what the data show. Identify the control group and the test group so you can see whether or not changes in the independent variable have had an effect. Look for differences in the dependent variable between the control and test groups.

These differences you observe can be qualitative or quantitative. You would be able to describe a qualitative difference using only words, whereas you would measure a quantitative difference and describe it using numbers. If there are differences, the independent variable that is being tested could have had an effect. If no differences are found between the control and test groups, the variable that is being tested apparently had no effect.

For example, suppose that three beakers each contain 100 mL of water. The beakers are placed on hot plates, and two of the hot plates are turned on, but the third is left off for a period of 5 min. Suppose you are then asked to describe any differences in the water in the three beakers. A qualitative difference might be the appearance of bubbles rising to the top in the water that is being heated but no rising bubbles in the unheated water. A quantitative difference might be a difference in the amount of water that is present in the beakers.

Inferring Scientists often make inferences based on their observations. An inference is an attempt to explain, or interpret, observations or to indicate what caused what you observed. An inference is a type of conclusion.

When making an inference, be certain to use accurate data and accurately described observations. Analyze all of the data that you've collected. Then, based on everything you know, explain or interpret what you've observed.

Drawing Conclusions

When scientists have analyzed the data they collected, they proceed to draw conclusions about what the data mean. These conclusions are sometimes stated using words similar to those found in the hypothesis formed earlier in the process.

Conclusions To analyze your data, you must review all of the observations and measurements that you made and recorded. Recheck all data for accuracy. After your data are rechecked and organized, you are almost ready to draw a conclusion such as "salt water boils at a higher temperature than freshwater."

Before you can draw a conclusion, however, you must determine whether the data allow you to come to a conclusion that supports a hypothesis. Sometimes that will be the case; other times it will not.

If your data do not support a hypothesis, it does not mean that the hypothesis is wrong. It means only that the results of the investigation did not support the hypothesis. Maybe the experiment needs to be redesigned, but very likely, some of the initial observations on which the hypothesis was based were incomplete or biased. Perhaps more observation or research is needed to refine the hypothesis.

Avoiding Bias Sometimes drawing a conclusion involves making judgments. When you make a judgment, you form an opinion about what your data mean. It is important to be honest and to avoid reaching a conclusion if there is no supporting evidence for it or if it is based on a small sample. It also is important not to allow any expectations of results to bias your judgments. If possible, it is a good idea to collect additional data. Scientists do this all the time.

For example, the *Hubble Space Telescope* was sent into space in April, 1990, to provide scientists with clearer views of the universe. *Hubble* is the size of a school bus and has a 2.4-m-diameter mirror. *Hubble* helped scientists answer questions about the planet Pluto.

For many years, scientists had only been able to hypothesize about the surface of the planet Pluto. *Hubble* has now provided pictures of Pluto's surface that show a rough texture with light and dark regions on it. This might be the best information about Pluto scientists will have until they are able to send a space probe to it.

Evaluating Others' Data and Conclusions

Sometimes scientists have to use data that they did not collect themselves, or they have to rely on observations and conclusions drawn by other researchers. In cases such as these, the data must be evaluated carefully.

How were the data obtained? How was the investigation done? Was it carried out properly? Has it been duplicated by other researchers? Were they able to follow the exact procedure? Did they come up with the same results? Look at the conclusion, as well. Would you reach the same conclusion from these results? Only when you have confidence in the data of others can you believe it is true and feel comfortable using it.

Communicating

The communication of ideas is an important part of the work of scientists. A discovery that is not reported will not advance the scientific community's understanding or knowledge. Communication among scientists also is important as a way of improving their investigations.

Scientists communicate in many ways, from writing articles in journals and magazines that explain their investigations and experiments, to announcing important discoveries on television and radio, to sharing ideas with colleagues on the Internet or presenting them as lectures.

People who study science rely on computers to record and store data and to analyze results from investigations. Whether you work in a laboratory or just need to write a lab report with tables, good computer skills are a necessity.

Using a Word Processor

Suppose your teacher has assigned a written report. After you've completed your research and decided how you want to write the information, you need to put all that information on paper. The easiest way to do this is with a word processing application on a computer.

A computer application that allows you to type your information, change it as many times as you need to, and then print it out so that it looks neat and clean is called a word processing application. You also can use this type of application to create tables and columns, add bullets or cartoon art to your page, include page numbers, and check your spelling.

Helpful Hints

- If you aren't sure how to do something using your word processing program, look in the help menu. You will find a list of topics there to click on for help. After you locate the help topic you need, just follow the step-by-step instructions you see on your screen.
- Just because you've spell checked your report doesn't mean that the spelling is perfect. The spell check feature can't catch misspelled words that look like other words. If you've accidentally typed *mind* instead of *mine*, the spell checker won't know the difference. Always reread your report to make sure you didn't miss any mistakes.

Figure 18
You can use computer programs to make graphs and tables.

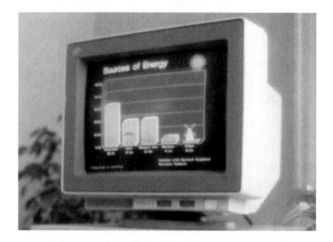

Using a Database

Imagine you're in the middle of a research project, busily gathering facts and information. You soon realize that it's becoming more difficult to organize and keep track of all the information. The tool to use to solve information overload is a database. Just as a file cabinet organizes paper records, a database organizes computer records. However, a database is more powerful than a simple file cabinet because at the click of a mouse, the contents can be reshuffled and reorganized. At computer-quick speeds, databases can sort information by any characteristics and filter data into multiple categories.

Helpful Hints

- Before setting up a database, take some time to learn the features of your database software by practicing with established database software.
- Periodically save your database as you enter data. That way, if something happens such as your computer malfunctions or the power goes off, you won't lose all of your work.

Doing a Database Search

When searching for information in a database, use the following search strategies to get the best results. These are the same search methods used for searching internet databases.

- Place the word *and* between two words in your search if you want the database to look for any entries that have both the words. For example, "Earth *and* Mars" would give you information that mentions both Earth and Mars.
- Place the word *or* between two words if you want the database to show entries that have at least one of the words. For example "Earth *or* Mars" would show you information that mentions either Earth or Mars.
- Place the word *not* between two words if you want the database to look for entries that have the first word but do not have the second word. For example, "Moon *not* phases" would show you information that mentions the Moon but does not mention its phases.

In summary, databases can be used to store large amounts of information about a particular subject. Databases allow biologists, Earth scientists, and physical scientists to search for information quickly and accurately.

Using an Electronic Spreadsheet

Your science fair experiment has produced lots of numbers. How do you keep track of all the data, and how can you easily work out all the calculations needed? You can use a computer program called a spreadsheet to record data that involve numbers. A spreadsheet is an electronic mathematical worksheet.

Type your data in rows and columns, just as they would look in a data table on a sheet of paper. A spreadsheet uses simple math to do data calculations. For example, you could add, subtract, divide, or multiply any of the values in the spreadsheet by another number. You also could set up a series of math steps you want to apply to the data. If you want to add 12 to all the numbers and then multiply all the numbers by 10, the computer does all the calculations for you in the spreadsheet. Below is an example of a spreadsheet that records weather data.

Helpful Hints

- Before you set up the spreadsheet, identify how you want to organize the data. Include any formulas you will need to use.
- Make sure you have entered the correct data into the correct rows and columns.
- You also can display your results in a graph. Pick the style of graph that best represents the data with which you are working.

Figure 19
A spreadsheet allows you to display large amounts of data and do calculations automatically.

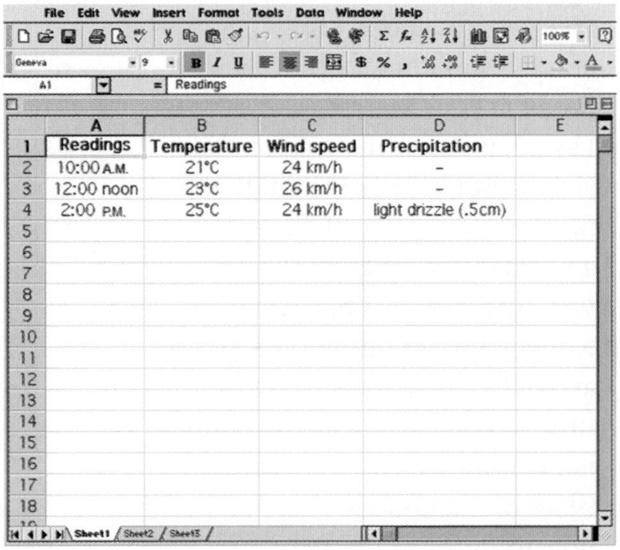

Using a Computerized Card Catalog

When you have a report or paper to research, you probably go to the library. To find the information you need in the library, you might have to use a computerized card catalog. This type of card catalog allows you to search for information by subject, by title, or by author. The computer then will display all the holdings the library has on the subject, title, or author requested.

A library's holdings can include books, magazines, databases, videos, and audio materials. When you have chosen something from this list, the computer will show whether an item is available and where in the library to find it.

Helpful Hints

- Remember that you can use the computer to search by subject, author, or title. If you know a book's author but not the title, you can search for all the books the library has by that author.
- When searching by subject, it's often most helpful to narrow your search by using specific search terms, such as *and, or,* and *not.* If you don't find enough sources, you can broaden your search.
- Pay attention to the type of materials found in your search. If you need a book, you can eliminate any videos or other resources that come up in your search.
- Knowing how your library is arranged can save you a lot of time. If you need help, the librarian will show you where certain types of materials are kept and how to find specific items.

Using Graphics Software

Are you having trouble finding that exact piece of art you're looking for? Do you have a picture in your mind of what you want but can't seem to find the right graphic to represent your ideas? To solve these problems, you can use graphics software. Graphics software allows you to create and change images and diagrams in almost unlimited ways. Typical uses for graphics software include arranging clip art, changing scanned images, and constructing pictures from scratch. Most graphics software applications work in similar ways. They use the same basic tools and functions. Once you master one graphics application, you can use other graphics applications.

Figure 20
Graphics software can use your data to draw bar graphs.

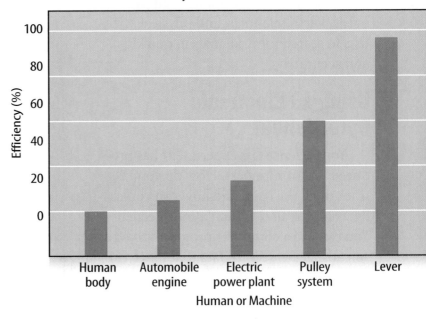

Efficiency of Humans and Machines

Figure 21
You can use this circle graph to find the names of the major gases that make up Earth's atmosphere.

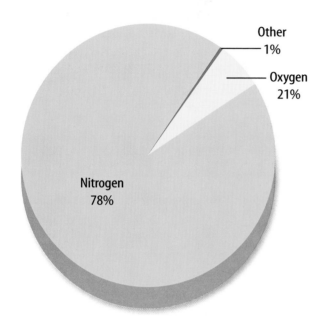

Helpful Hints

- As with any method of drawing, the more you practice using the graphics software, the better your results will be.
- Start by using the software to manipulate existing drawings. Once you master this, making your own illustrations will be easier.
- Clip art is available on CD-ROMs and the Internet. With these resources, finding a piece of clip art to suit your purposes is simple.
- As you work on a drawing, save it often.

Developing Multimedia Presentations

It's your turn—you have to present your science report to the entire class. How do you do it? You can use many different sources of information to get the class excited about your presentation. Posters, videos, photographs, sound, computers, and the Internet can help show your ideas.

First, determine what important points you want to make in your presentation. Then, write an outline of what materials and types of media would best illustrate those points. Maybe you could start with an outline on an overhead projector, then show a video, followed by something from the Internet or a slide show accompanied by music or recorded voices. You might choose to use a presentation builder computer application that can combine all these elements into one presentation. Make sure the presentation is well constructed to make the most impact on the audience.

Figure 22
Multimedia presentations use many types of print and electronic materials.

Helpful Hints

- Carefully consider what media will best communicate the point you are trying to make.
- Make sure you know how to use any equipment you will be using in your presentation.
- Practice the presentation several times.
- If possible, set up all of the equipment ahead of time. Make sure everything is working correctly.

Math Skill Handbook

Use this Math Skill Handbook to help solve problems you are given in this text. You might find it useful to review topics in this Math Skill Handbook first.

Skill Handbooks

Converting Units

In science, quantities such as length, mass, and time sometimes are measured using different units. Suppose you want to know how many miles are in 12.7 km.

Conversion factors are used to change from one unit of measure to another. A conversion factor is a ratio that is equal to one. For example, there are 1,000 mL in 1 L, so 1,000 mL equals 1 L, or:

$$1{,}000 \text{ mL} = 1 \text{ L}$$

If both sides are divided by 1 L, this equation becomes:

$$\frac{1{,}000 \text{ mL}}{1 \text{ L}} = 1$$

The **ratio** on the left side of this equation is equal to 1 and is a conversion factor. You can make another conversion factor by dividing both sides of the top equation by 1,000 mL:

$$1 = \frac{1 \text{ L}}{1{,}000 \text{ mL}}$$

To **convert units,** you multiply by the appropriate conversion factor. For example, how many milliliters are in 1.255 L? To convert 1.255 L to milliliters, multiply 1.255 L by a conversion factor.

Use the **conversion factor** with new units (mL) in the numerator and the old units (L) in the denominator.

$$1.255 \text{ L} \times \frac{1{,}000 \text{ mL}}{1 \text{ L}} = 1{,}255 \text{ mL}$$

The unit L divides in this equation, just as if it were a number.

Example 1   There are 2.54 cm in 1 inch. If a meterstick has a length of 100 cm, how long is the meterstick in inches?

Step 1   Decide which conversion factor to use. You know the length of the meterstick in centimeters, so centimeters are the old units. You want to find the length in inches, so inch is the new unit.

Step 2   Form the conversion factor. Start with the relationship between the old and new units.

$$2.54 \text{ cm} = 1 \text{ inch}$$

Step 3   Form the conversion factor with the old unit (centimeter) on the bottom by dividing both sides by 2.54 cm.

$$1 = \frac{2.54 \text{ cm}}{2.54 \text{ cm}} = \frac{1 \text{ inch}}{2.54 \text{ cm}}$$

Step 4   Multiply the old measurement by the conversion factor.

$$100 \text{ cm} \times \frac{1 \text{ inch}}{2.54 \text{ cm}} = 39.37 \text{ inches}$$

The meterstick is 39.37 inches long.

Example 2   There are 365 days in one year. If a person is 14 years old, what is his or her age in days? (Ignore leap years).

Step 1   Decide which conversion factor to use. You want to convert years to days.

Step 2   Form the conversion factor. Start with the relation between the old and new units.

$$1 \text{ year} = 365 \text{ days}$$

Step 3   Form the conversion factor with the old unit (year) on the bottom by dividing both sides by 1 year.

$$1 = \frac{1 \text{ year}}{1 \text{ year}} = \frac{365 \text{ days}}{1 \text{ year}}$$

Step 4   Multiply the old measurement by the conversion factor:

$$14 \text{ years} \times \frac{365 \text{ days}}{1 \text{ year}} = 5{,}110 \text{ days}$$

The person's age is 5,110 days.

Practice Problem   A book has a mass of 2.31 kg. If there are 1,000 g in 1 kg, what is the mass of the book in grams?

Using Fractions

A **fraction** is a number that compares a part to the whole. For example, in the fraction $\frac{2}{3}$, the 2 represents the part and the 3 represents the whole. In the fraction $\frac{2}{3}$, the top number, 2, is called the numerator. The bottom number, 3, is called the denominator.

Sometimes fractions are not written in their simplest form. To determine a fraction's **simplest form,** you must find the greatest common factor (GCF) of the numerator and denominator. The greatest common factor is the largest common factor of all the factors the two numbers have in common.

For example, because the number 3 divides into 12 and 30 evenly, it is a common factor of 12 and 30. However, because the number 6 is the largest number that evenly divides into 12 and 30, it is the **greatest common factor.**

After you find the greatest common factor, you can write a fraction in its simplest form. Divide both the numerator and the denominator by the greatest common factor. The number that results is the fraction in its **simplest form.**

Example Twelve of the 20 peaks in a mountain range have elevations over 10,000 m. What fraction of the peaks in the mountain range are over 10,000 m? Write the fraction in simplest form.

Step 1 Write the fraction.

$$\frac{part}{whole} = \frac{12}{20}$$

Step 2 To find the GCF of the numerator and denominator, list all of the factors of each number.

Factors of 12: 1, 2, 3, 4, 6, 12 (the numbers that divide evenly into 12)

Factors of 20: 1, 2, 4, 5, 10, 20 (the numbers that divide evenly into 20)

Step 3 List the common factors.

1, 2, 4.

Step 4 Choose the greatest factor in the list of common factors.

The GCF of 12 and 20 is 4.

Step 5 Divide the numerator and denominator by the GCF.

$$\frac{12 \div 4}{20 \div 4} = \frac{3}{5}$$

In the mountain range, $\frac{3}{5}$ of the peaks are over 10,000 m.

Practice Problem There are 90 rides at an amusement park. Of those rides, 66 have a height restriction. What fraction of the rides has a height restriction? Write the fraction in simplest form.

Math Skill Handbook

Skill Handbooks

Calculating Ratios

A **ratio** is a comparison of two numbers by division.

Ratios can be written 3 to 5 or 3:5. Ratios also can be written as fractions, such as $\frac{3}{5}$. Ratios, like fractions, can be written in simplest form. Recall that a fraction is in **simplest form** when the greatest common factor (GCF) of the numerator and denominator is 1.

Example A particular geologic sample contains 40 kg of shale and 64 kg of granite. What is the ratio of shale to granite as a fraction in simplest form?

Step 1 Write the ratio as a fraction. $\frac{shale}{granite} = \frac{40}{64}$

Step 2 Express the fraction in simplest form. The GCF of 40 and 64 is 8.

$$\frac{40}{64} = \frac{40 \div 8}{64 \div 8} = \frac{5}{8}$$

The ratio of shale to granite in the sample is $\frac{5}{8}$.

Practice Problem Two metal rods measure 100 cm and 144 cm in length. What is the ratio of their lengths in simplest fraction form?

Using Decimals

A **decimal** is a fraction with a denominator of 10, 100, 1,000, or another power of 10. For example, 0.854 is the same as the fraction $\frac{854}{1,000}$.

In a decimal, the decimal point separates the ones place and the tenths place. For example, 0.27 means twenty-seven hundredths, or $\frac{27}{100}$, where 27 is the **number of units** out of 100 units. Any fraction can be written as a decimal using division.

Example Write $\frac{5}{8}$ as a decimal.

Step 1 Write a division problem with the numerator, 5, as the dividend and the denominator, 8, as the divisor. Write 5 as 5.000.

Step 2 Solve the problem.

```
      0.625
   8)5.000
      48
      20
      16
       40
       40
        0
```

Therefore, $\frac{5}{8} = 0.625$.

Practice Problem Write $\frac{19}{25}$ as a decimal.

Using Percentages

The word *percent* means "out of one hundred." A **percent** is a ratio that compares a number to 100. Suppose you read that 77 percent of Earth's surface is covered by water. That is the same as reading that the fraction of Earth's surface covered by water is $\frac{77}{100}$. To express a fraction as a percent, first find an equivalent decimal for the fraction. Then, multiply the decimal by 100 and add the percent symbol. For example, $\frac{1}{2} = 1 \div 2 = 0.5$. Then $0.5 \cdot 100 = 50 = 50\%$.

Example Express $\frac{13}{20}$ as a percent.

Step 1 Find the equivalent decimal for the fraction.

$$
\begin{array}{r}
0.65 \\
20\overline{)13.00} \\
\underline{120} \\
100 \\
\underline{100} \\
0
\end{array}
$$

Step 2 Rewrite the fraction $\frac{13}{20}$ as 0.65.

Step 3 Multiply 0.65 by 100 and add the % sign.

$0.65 \cdot 100 = 65 = 65\%$

So, $\frac{13}{20} = 65\%$.

Practice Problem In one year, 73 of 365 days were rainy in one city. What percent of the days in that city were rainy?

Using Precision and Significant Digits

When you make a **measurement,** the value you record depends on the precision of the measuring instrument. When adding or subtracting numbers with different precision, the answer is rounded to the smallest number of decimal places of any number in the sum or difference. When multiplying or dividing, the answer is rounded to the smallest number of significant figures of any number being multiplied or divided. When counting the number of **significant figures,** all digits are counted except zeros at the end of a number with no decimal such as 2,500, and zeros at the beginning of a decimal such as 0.03020.

Example The lengths 5.28 and 5.2 are measured in meters. Find the sum of these lengths and report the sum using the least precise measurement.

Step 1 Find the sum.

$$
\begin{array}{ll}
5.28 \text{ m} & \text{2 digits after the decimal} \\
+\ 5.2 \text{ m} & \text{1 digit after the decimal} \\
\hline
10.48 \text{ m} &
\end{array}
$$

Step 2 Round to one digit after the decimal because the least number of digits after the decimal of the numbers being added is 1.

The sum is 10.5 m.

Practice Problem Multiply the numbers in the example using the rule for multiplying and dividing. Report the answer with the correct number of significant figures.

Math Skill Handbook

Solving One-Step Equations

An **equation** is a statement that two things are equal. For example, $A = B$ is an equation that states that A is equal to B.

Sometimes one side of the equation will contain a **variable** whose value is not known. In the equation $3x = 12$, the variable is x.

The equation is solved when the variable is replaced with a value that makes both sides of the equation equal to each other. For example, the solution of the equation $3x = 12$ is $x = 4$. If the x is replaced with 4, then the equation becomes $3 \cdot 4 = 12$, or $12 = 12$.

To solve an equation such as $8x = 40$, divide both sides of the equation by the number that multiplies the variable.

$$8x = 40$$
$$\frac{8x}{8} = \frac{40}{8}$$
$$x = 5$$

You can check your answer by replacing the variable with your solution and seeing if both sides of the equation are the same.

$$8x = 8 \cdot 5 = 40$$

The left and right sides of the equation are the same, so $x = 5$ is the solution.

Sometimes an equation is written in this way: $a = bc$. This also is called a **formula.** The letters can be replaced by numbers, but the numbers must still make both sides of the equation the same.

Example 1 Solve the equation $10x = 35$.

Step 1 Find the solution by dividing each side of the equation by 10.

$$10x = 35 \qquad \frac{10x}{10} = \frac{35}{10} \qquad x = 3.5$$

Step 2 Check the solution.

$$10x = 35 \qquad 10 \times 3.5 = 35 \qquad 35 = 35$$

Both sides of the equation are equal, so $x = 3.5$ is the solution to the equation.

Example 2 In the formula $a = bc$, find the value of c if $a = 20$ and $b = 2$.

Step 1 Rearrange the formula so the unknown value is by itself on one side of the equation by dividing both sides by b.

$$a = bc$$
$$\frac{a}{b} = \frac{bc}{b}$$
$$\frac{a}{b} = c$$

Step 2 Replace the variables a and b with the values that are given.

$$\frac{a}{b} = c$$
$$\frac{20}{2} = c$$
$$10 = c$$

Step 3 Check the solution.

$$a = bc$$
$$20 = 2 \times 10$$
$$20 = 20$$

Both sides of the equation are equal, so $c = 10$ is the solution when $a = 20$ and $b = 2$.

Practice Problem In the formula $h = gd$, find the value of d if $g = 12.3$ and $h = 17.4$.

A **proportion** is an equation that shows that two ratios are equivalent. The ratios $\frac{2}{4}$ and $\frac{5}{10}$ are equivalent, so they can be written as $\frac{2}{4} = \frac{5}{10}$. This equation is an example of a proportion.

When two ratios form a proportion, the **cross products** are equal. To find the cross products in the proportion $\frac{2}{4} = \frac{5}{10}$, multiply the 2 and the 10, and the 4 and the 5. Therefore $2 \cdot 10 = 4 \cdot 5$, or $20 = 20$.

Because you know that both proportions are equal, you can use cross products to find a missing term in a proportion. This is known as **solving the proportion.** Solving a proportion is similar to solving an equation.

Example The heights of a tree and a pole are proportional to the lengths of their shadows. The tree casts a shadow of 24 m at the same time that a 6-m pole casts a shadow of 4 m. What is the height of the tree?

Step 1 Write a proportion.

$$\frac{\text{height of tree}}{\text{height of pole}} = \frac{\text{length of tree's shadow}}{\text{length of pole's shadow}}$$

Step 2 Substitute the known values into the proportion. Let h represent the unknown value, the height of the tree.

$$\frac{h}{6} = \frac{24}{4}$$

Step 3 Find the cross products.

$$h \cdot 4 = 6 \cdot 24$$

Step 4 Simplify the equation.

$$4h = 144$$

Step 5 Divide each side by 4.

$$\frac{4h}{4} = \frac{144}{4}$$

$$h = 36$$

The height of the tree is 36 m.

Practice Problem The ratios of the weights of two objects on the Moon and on Earth are in proportion. A rock weighing 3 N on the Moon weighs 18 N on Earth. How much would a rock that weighs 5 N on the Moon weigh on Earth?

Math Skill Handbook

Using Statistics

Statistics is the branch of mathematics that deals with collecting, analyzing, and presenting data. In statistics, there are three common ways to summarize the data with a single number—the mean, the median, and the mode.

The **mean** of a set of data is the arithmetic average. It is found by adding the numbers in the data set and dividing by the number of items in the set.

The **median** is the middle number in a set of data when the data are arranged in numerical order. If there were an even number of data points, the median would be the mean of the two middle numbers.

The **mode** of a set of data is the number or item that appears most often.

Another number that often is used to describe a set of data is the range. The **range** is the difference between the largest number and the smallest number in a set of data.

A **frequency table** shows how many times each piece of data occurs, usually in a survey. The frequency table below shows the results of a student survey on favorite color.

Color	Tally	Frequency
red	||||	4
blue	⊬⊬	5
black	||	2
green	|||	3
purple	⊬⊬ ||	7
yellow	⊬⊬ |	6

Based on the frequency table data, which color is the favorite?

Example The high temperatures (in °C) on five consecutive days at a desert observation station are 39°, 37°, 44°, 36°, and 44°. Find the mean, median, mode, and range of this set.

To find the mean:

Step 1 Find the sum of the numbers.

$$39 + 37 + 44 + 36 + 44 = 200$$

Step 2 Divide the sum by the number of items, which is 5.

$$200 \div 5 = 40$$

The mean high temperature is 40°C.

To find the median:

Step 1 Arrange the temperatures from least to greatest.

$$36, \ 37, \ \underline{39}, \ 44, \ 44$$

Step 2 Determine the middle temperature.

The median high temperature is 39°C.

To find the mode:

Step 1 Group the numbers that are the same together.

$$44, 44, 36, 37, 39$$

Step 2 Determine the number that occurs most in the set.

$$\underline{44, 44}, 36, 37, 39$$

The mode measure is 44°C.

To find the range:

Step 1 Arrange the temperatures from largest to smallest.

$$44, 44, 39, 37, 36$$

Step 2 Determine the largest and smallest temperature in the set.

$$\underline{44}, 44, 39, 37, \underline{36}$$

Step 3 Find the difference between the largest and smallest temperatures.

$$44 - 36 = 8$$

The range is 8°C.

Practice Problem Find the mean, median, mode, and range for the data set 8, 4, 12, 8, 11, 14, 16.

Safety in the Science Classroom

1. Always obtain your teacher's permission to begin an investigation.

2. Study the procedure. If you have questions, ask your teacher. Be sure you understand any safety symbols shown on the page.

3. Use the safety equipment provided for you. Goggles and a safety apron should be worn during most investigations.

4. Always slant test tubes away from yourself and others when heating them or adding substances to them.

5. Never eat or drink in the lab, and never use lab glassware as food or drink containers. Never inhale chemicals. Do not taste any substances or draw any material into a tube with your mouth.

6. Report any spill, accident, or injury, no matter how small, immediately to your teacher; then follow his or her instructions.

7. Know the location and proper use of the fire extinguisher, safety shower, fire blanket, first aid kit, and fire alarm.

8. Keep all materials away from open flames. Tie back long hair and tie down loose clothing.

9. If your clothing should catch fire, smother it with the fire blanket, or get under a safety shower. NEVER RUN.

10. If a fire should occur, turn off the gas; then leave the room according to established procedures.

Follow these procedures as you clean up your work area

1. Turn off the water and gas. Disconnect electrical devices.

2. Clean all pieces of equipment and return all materials to their proper places.

3. Dispose of chemicals and other materials as directed by your teacher. Place broken glass and solid substances in the proper containers. Make sure never to discard materials in the sink.

4. Clean your work area. Wash your hands thoroughly after working in the laboratory.

First Aid	
Injury	**Safe Response ALWAYS NOTIFY YOUR TEACHER IMMEDIATELY**
Burns	Apply cold water.
Cuts and Bruises	Stop any bleeding by applying direct pressure. Cover cuts with a clean dressing. Apply ice packs or cold compresses to bruises.
Fainting	Leave the person lying down. Loosen any tight clothing and keep crowds away.
Foreign Matter in Eye	Flush with plenty of water. Use eyewash bottle or fountain.
Poisoning	Note the suspected poisoning agent.
Any Spills on Skin	Flush with large amounts of water or use safety shower.

PERIODIC TABLE OF THE ELEMENTS

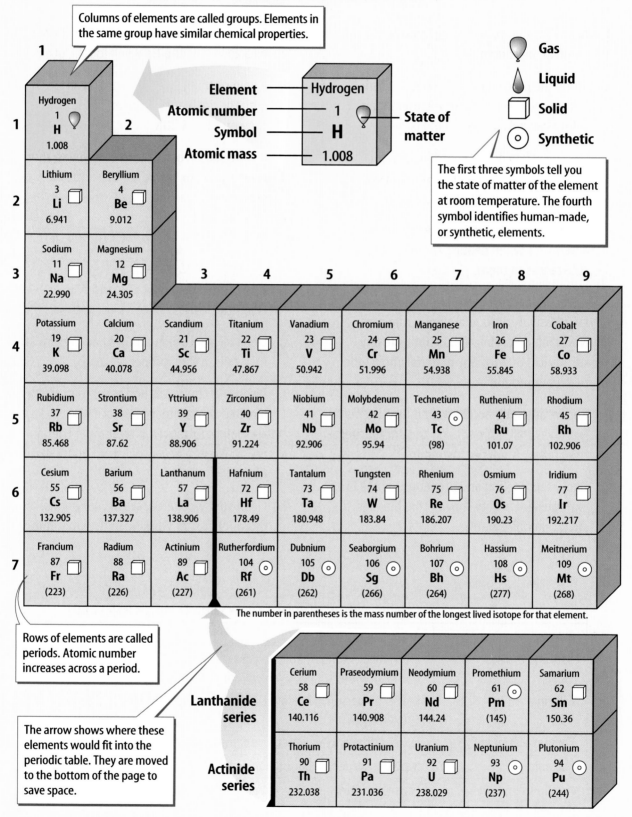

Columns of elements are called groups. Elements in the same group have similar chemical properties.

Element ——— Hydrogen
Atomic number ——— 1
Symbol ——— **H**
Atomic mass ——— 1.008

State of matter

Gas
Liquid
Solid
Synthetic

The first three symbols tell you the state of matter of the element at room temperature. The fourth symbol identifies human-made, or synthetic, elements.

1								
Hydrogen 1 **H** 1.008	2							
Lithium 3 **Li** 6.941	Beryllium 4 **Be** 9.012							
Sodium 11 **Na** 22.990	Magnesium 12 **Mg** 24.305	3	4	5	6	7	8	9
Potassium 19 **K** 39.098	Calcium 20 **Ca** 40.078	Scandium 21 **Sc** 44.956	Titanium 22 **Ti** 47.867	Vanadium 23 **V** 50.942	Chromium 24 **Cr** 51.996	Manganese 25 **Mn** 54.938	Iron 26 **Fe** 55.845	Cobalt 27 **Co** 58.933
Rubidium 37 **Rb** 85.468	Strontium 38 **Sr** 87.62	Yttrium 39 **Y** 88.906	Zirconium 40 **Zr** 91.224	Niobium 41 **Nb** 92.906	Molybdenum 42 **Mo** 95.94	Technetium 43 **Tc** (98)	Ruthenium 44 **Ru** 101.07	Rhodium 45 **Rh** 102.906
Cesium 55 **Cs** 132.905	Barium 56 **Ba** 137.327	Lanthanum 57 **La** 138.906	Hafnium 72 **Hf** 178.49	Tantalum 73 **Ta** 180.948	Tungsten 74 **W** 183.84	Rhenium 75 **Re** 186.207	Osmium 76 **Os** 190.23	Iridium 77 **Ir** 192.217
Francium 87 **Fr** (223)	Radium 88 **Ra** (226)	Actinium 89 **Ac** (227)	Rutherfordium 104 **Rf** (261)	Dubnium 105 **Db** (262)	Seaborgium 106 **Sg** (266)	Bohrium 107 **Bh** (264)	Hassium 108 **Hs** (277)	Meitnerium 109 **Mt** (268)

The number in parentheses is the mass number of the longest lived isotope for that element.

Rows of elements are called periods. Atomic number increases across a period.

The arrow shows where these elements would fit into the periodic table. They are moved to the bottom of the page to save space.

	Lanthanide series	Cerium 58 **Ce** 140.116	Praseodymium 59 **Pr** 140.908	Neodymium 60 **Nd** 144.24	Promethium 61 **Pm** (145)	Samarium 62 **Sm** 150.36
	Actinide series	Thorium 90 **Th** 232.038	Protactinium 91 **Pa** 231.036	Uranium 92 **U** 238.029	Neptunium 93 **Np** (237)	Plutonium 94 **Pu** (244)

Reference Handbook

REFERENCE HANDBOOK B

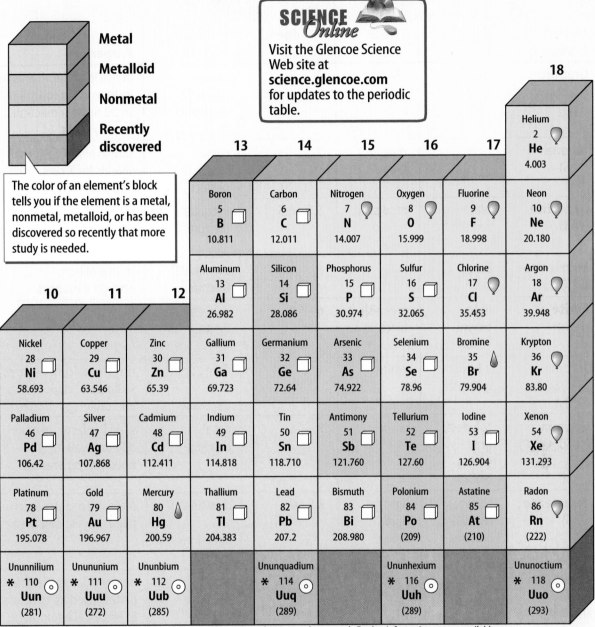

Metal

Metalloid

Nonmetal

Recently discovered

The color of an element's block tells you if the element is a metal, nonmetal, metalloid, or has been discovered so recently that more study is needed.

SCIENCE Online

Visit the Glencoe Science Web site at **science.glencoe.com** for updates to the periodic table.

		13	14	15	16	17	18
							Helium 2 He 4.003
		Boron 5 B 10.811	Carbon 6 C 12.011	Nitrogen 7 N 14.007	Oxygen 8 O 15.999	Fluorine 9 F 18.998	Neon 10 Ne 20.180
10	11	12 Aluminum 13 Al 26.982	Silicon 14 Si 28.086	Phosphorus 15 P 30.974	Sulfur 16 S 32.065	Chlorine 17 Cl 35.453	Argon 18 Ar 39.948
Nickel 28 Ni 58.693	Copper 29 Cu 63.546	Zinc 30 Zn 65.39	Gallium 31 Ga 69.723	Germanium 32 Ge 72.64	Arsenic 33 As 74.922	Selenium 34 Se 78.96	Bromine 35 Br 79.904	Krypton 36 Kr 83.80
Palladium 46 Pd 106.42	Silver 47 Ag 107.868	Cadmium 48 Cd 112.411	Indium 49 In 114.818	Tin 50 Sn 118.710	Antimony 51 Sb 121.760	Tellurium 52 Te 127.60	Iodine 53 I 126.904	Xenon 54 Xe 131.293
Platinum 78 Pt 195.078	Gold 79 Au 196.967	Mercury 80 Hg 200.59	Thallium 81 Tl 204.383	Lead 82 Pb 207.2	Bismuth 83 Bi 208.980	Polonium 84 Po (209)	Astatine 85 At (210)	Radon 86 Rn (222)
Ununnilium * 110 Uun (281)	Unununium * 111 Uuu (272)	Ununbium * 112 Uub (285)		Ununquadium * 114 Uuq (289)		Ununhexium * 116 Uuh (289)		Ununoctium * 118 Uuo (293)

* Names not officially assigned. Discovery of elements 114, 116, and 118 recently reported. Further information not yet available.

Europium 63 Eu 151.964	Gadolinium 64 Gd 157.25	Terbium 65 Tb 158.925	Dysprosium 66 Dy 162.50	Holmium 67 Ho 164.930	Erbium 68 Er 167.259	Thulium 69 Tm 168.934	Ytterbium 70 Yb 173.04	Lutetium 71 Lu 174.967
Americium 95 Am (243)	Curium 96 Cm (247)	Berkelium 97 Bk (247)	Californium 98 Cf (251)	Einsteinium 99 Es (252)	Fermium 100 Fm (257)	Mendelevium 101 Md (258)	Nobelium 102 No (259)	Lawrencium 103 Lr (262)

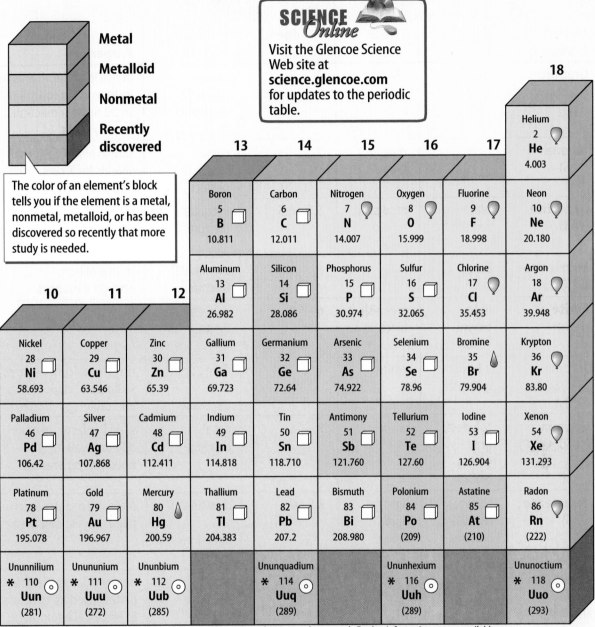

REFERENCE HANDBOOK C

Minerals

Mineral (formula)	Color	Streak	Hardness	Breakage Pattern	Uses and Other Properties
Graphite (C)	black to gray	black to gray	1–1.5	basal cleavage (scales)	pencil lead, lubricants for locks, rods to control some small nuclear reactions, battery poles
Galena (PbS)	gray	gray to black	2.5	cubic cleavage perfect	source of lead, used for pipes, shields for X rays, fishing equipment sinkers
Hematite (Fe_2O_3)	black or reddish-brown	reddish-brown	5.5–6.5	irregular fracture	source of iron; converted to pig iron, made into steel
Magnetite (Fe_3O_4)	black	black	6	conchoidal fracture	source of iron, attracts a magnet
Pyrite (FeS_2)	light, brassy, yellow	greenish-black	6–6.5	uneven fracture	fool's gold
Talc ($Mg_3 Si_4O_{10}$ $(OH)_2$)	white, greenish	white	1	cleavage in one direction	used for talcum powder, sculptures, paper, and tabletops
Gypsum ($CaSO_4 \cdot 2H_2O$)	colorless, gray, white, brown	white	2	basal cleavage	used in plaster of paris and dry wall for building construction
Sphalerite (ZnS)	brown, reddish-brown, greenish	light to dark brown	3.5–4	cleavage in six directions	main ore of zinc; used in paints, dyes, and medicine
Muscovite (KAl_3Si_3 $O_{10}(OH)_2$)	white, light gray, yellow, rose, green	colorless	2–2.5	basal cleavage	occurs in large, flexible plates; used as an insulator in electrical equipment, lubricant
Biotite ($K(Mg,Fe)_3$ $(AlSi_3O_{10})$ $(OH)_2$)	black to dark brown	colorless	2.5–3	basal cleavage	occurs in large, flexible plates
Halite (NaCl)	colorless, red, white, blue	colorless	2.5	cubic cleavage	salt; soluble in water; a preservative

REFERENCE HANDBOOK C

Minerals

Mineral (formula)	Color	Streak	Hardness	Breakage Pattern	Uses and Other Properties
Calcite ($CaCO_3$)	colorless, white, pale blue	colorless, white	3	cleavage in three directions	fizzes when HCl is added; used in cements and other building materials
Dolomite ($CaMg(CO_3)_2$)	colorless, white, pink, green, gray, black	white	3.5–4	cleavage in three directions	concrete and cement; used as an ornamental building stone
Fluorite (CaF_2)	colorless, white, blue, green, red, yellow, purple	colorless	4	cleavage in four directions	used in the manufacture of optical equipment; glows under ultraviolet light
Hornblende ($(CaNa)_{2-3}$ $(Mg,Al,$ $Fe)_5-(Al,Si)_2$ Si_6O_{22} $(OH)_2)$	green to black	gray to white	5–6	cleavage in two directions	will transmit light on thin edges; 6-sided cross section
Feldspar ($KAlSi_3O_8$) ($NaAl$ Si_3O_8), ($CaAl_2Si_2$ O_8)	colorless, white to gray, green	colorless	6	two cleavage planes meet at 90° angle	used in the manufacture of ceramics
Augite ((Ca,Na) (Mg,Fe,Al) $(Al,Si)_2 O_6$)	black	colorless	6	cleavage in two directions	square or 8-sided cross section
Olivine ($(Mg,Fe)_2$ SiO_4)	olive, green	none	6.5–7	conchoidal fracture	gemstones, refractory sand
Quartz (SiO_2)	colorless, various colors	none	7	conchoidal fracture	used in glass manufacture, electronic equipment, radios, computers, watches, gemstones

Reference Handbook

REFERENCE HANDBOOK D

Rocks

Rock Type	Rock Name	Characteristics
Igneous (intrusive)	Granite	Large mineral grains of quartz, feldspar, hornblende, and mica. Usually light in color.
	Diorite	Large mineral grains of feldspar, hornblende, and mica. Less quartz than granite. Intermediate in color.
	Gabbro	Large mineral grains of feldspar, augite, and olivine. No quartz. Dark in color.
Igneous (extrusive)	Rhyolite	Small mineral grains of quartz, feldspar, hornblende, and mica, or no visible grains. Light in color.
	Andesite	Small mineral grains of feldspar, hornblende, and mica or no visible grains. Intermediate in color.
	Basalt	Small mineral grains of feldspar, augite, and olivine or no visible grains. No quartz. Dark in color.
	Obsidian	Glassy texture. No visible grains. Volcanic glass. Fracture looks like broken glass.
	Pumice	Frothy texture. Floats in water. Usually light in color.
Sedimentary (detrital)	Conglomerate	Coarse grained. Gravel or pebble size grains.
	Sandstone	Sand-sized grains 1/16 to 2 mm.
	Siltstone	Grains are smaller than sand but larger than clay.
	Shale	Smallest grains. Often dark in color. Usually platy.
Sedimentary (chemical or organic)	Limestone	Major mineral is calcite. Usually forms in oceans, lakes, and caves. Often contains fossils.
	Coal	Occurs in swampy areas. Compacted layers of organic material, mainly plant remains.
Sedimentary (chemical)	Rock Salt	Commonly forms by the evaporation of seawater.
Metamorphic (foliated)	Gneiss	Banding due to alternate layers of different minerals, of different colors. Parent rock often is granite.
	Schist	Parallel arrangement of sheetlike minerals, mainly micas. Forms from different parent rocks.
	Phyllite	Shiny or silky appearance. May look wrinkled. Common parent rocks are shale and slate.
	Slate	Harder, denser, and shinier than shale. Common parent rock is shale.
Metamorphic (non-foliated)	Marble	Calcite or dolomite. Common parent rock is limestone.
	Soapstone	Mainly of talc. Soft with greasy feel.
	Quartzite	Hard with interlocking quartz crystals. Common parent rock is sandstone.

SI—Metric/English, English/Metric Conversions

	When you want to convert:	To:	Multiply by:
Length	inches	centimeters	2.54
	centimeters	inches	0.39
	yards	meters	0.91
	meters	yards	1.09
	miles	kilometers	1.61
	kilometers	miles	0.62
Mass and Weight*	ounces	grams	28.35
	grams	ounces	0.04
	pounds	kilograms	0.45
	kilograms	pounds	2.2
	tons (short)	tonnes (metric tons)	0.91
	tonnes (metric tons)	tons (short)	1.10
	pounds	newtons	4.45
	newtons	pounds	0.22
Volume	cubic inches	cubic centimeters	16.39
	cubic centimeters	cubic inches	0.06
	liters	quarts	1.06
	quarts	liters	0.95
	gallons	liters	3.78
Area	square inches	square centimeters	6.45
	square centimeters	square inches	0.16
	square yards	square meters	0.83
	square meters	square yards	1.19
	square miles	square kilometers	2.59
	square kilometers	square miles	0.39
	hectares	acres	2.47
	acres	hectares	0.40
Temperature	To convert °Celsius to °Fahrenheit		$°C \times 9/5 + 32$
	To convert °Fahrenheit to °Celsius		$5/9 \, (°F - 32)$

*Weight is measured in standard Earth gravity.

English Glossary

This glossary defines each key term that appears in bold type in the text. It also shows the chapter, section, and page number where you can find the word used.

asthenosphere (as THE nuh sfihr): plasticlike layer of Earth on which the lithospheric plates float and move around. (Chap. 4, Sec. 3, p. 106)

B

basaltic: dense, dark-colored igneous rock formed from magma rich in magnesium and iron and poor in silica. (Chap. 2, Sec. 2, p. 43)

batholith: largest intrusive igneous rock body that forms when magma being forced upward toward Earth's crust cools slowly and solidifies underground. (Chap. 6, Sec. 3, p. 172)

biomass energy: renewable energy derived from burning organic materials such as wood and alcohol. (Chap. 3, Sec. 2, p. 79)

C

caldera: large, circular-shaped opening formed when the top of a volcano collapses. (Chap. 6, Sec. 3, p. 174)

cementation: sedimentary rock-forming process in which large sediments are held together by natural cements that are produced when water soaks through rock and soil. (Chap. 2, Sec. 4, p. 51)

cinder cone volcano: steep-sided, loosely packed volcano formed when tephra falls to the ground. (Chap. 6, Sec. 2, p. 166)

cleavage: physical property of some minerals that causes them to break along smooth, flat surfaces. (Chap. 1, Sec. 2, p. 17)

coal: sedimentary rock formed from decayed plant material; the world's most abundant fossil fuel. (Chap. 3, Sec. 1, p. 67)

compaction: process that forms sedimentary rocks when layers of small sediments are compressed by the weight of the layers above them. (Chap. 2, Sec. 4, p. 50)

composite volcano: volcano built by alternating explosive and quiet eruptions that produce layers of tephra and lava; found mostly where Earth's plates come together and one plate sinks below the other. (Chap. 6, Sec. 2, p. 167)

continental drift: Wegener's hypothesis that all continents were once connected in a single large landmass that broke apart about 200 million years ago and drifted slowly to their current positions. (Chap. 4 Sec. 1, p. 98)

convection current: current in Earth's mantle that transfers heat in Earth's interior and is the driving force for plate tectonics. (Chap. 4, Sec. 3, p. 111)

crater: steep-walled depression around a volcano's vent. (Chap. 6, Sec. 1, p. 158)

crystal: solid in which the atoms are arranged in an orderly, repeating pattern. (Chap. 1, Sec. 1, p. 9)

D

dike: igneous rock feature formed when magma is squeezed into a vertical crack that cuts across rock layers and hardens underground. (Chap. 6, Sec. 3, p. 173)

E

earthquake: vibrations produced when rocks break along a fault. (Chap. 5, Sec. 1, p. 127)

epicenter (EP ih sent ur): point on Earth's surface directly above an earthquake's focus. (Chap. 5, Sec. 2, p. 131)

extrusive: fine-grained igneous rock that forms when magma cools quickly at or near Earth's surface. (Chap. 2, Sec. 2, p. 41)

F

fault: surface along which rocks move when they pass their elastic limit and break. (Chap. 5, Sec. 1, p. 126)

focus: in an earthquake, the point below Earth's surface where energy is released in the form of seismic waves. (Chap. 5, Sec. 2, p. 130)

foliated: metamorphic rock, such as slate and gneiss, whose mineral grains flatten and line up in parallel layers. (Chap. 2, Sec. 3, p. 47)

fossil fuel: nonrenewable energy resource, such as oil and coal, formed over millions of years from the remains of dead plants and other organisms. (Chap. 3, Sec. 1, p. 66)

fracture: physical property of some minerals that causes them to break with uneven, rough, or jagged surfaces. (Chap. 1, Sec. 2, p. 17)

G

gem: beautiful, rare, highly prized mineral that can be worn in jewelry. (Chap. 1, Sec. 3, p. 19)

geothermal energy: inexhaustible energy resource that uses hot magma or hot, dry rocks from below Earth's surface to generate electricity. (Chap. 3, Sec. 2, p. 78)

granitic: light-colored, silica-rich igneous rock that is less dense than basaltic rock. (Chap. 2, Sec. 2, p. 43)

H

hardness: measure of how easily a mineral can be scratched; is determined by the arrangement of the mineral's atoms. (Chap. 1, Sec. 2, p. 15)

hot spot: unusually hot area at the boundary between Earth's mantle and core that forms volcanoes when melted rock is forced upward and breaks through the crust. (Chap. 6, Sec. 1, p. 160)

hydroelectric energy: electricity produced by waterpower using large dams in a river. (Chap. 3, Sec. 2, p. 78)

I

igneous rock: intrusive or extrusive rock formed when hot magma cools and hardens. (Chap. 2, Sec. 2, p. 40)

English Glossary

intrusive: a type of igneous rock that generally contains large crystals and forms when magma cools slowly beneath Earth's surface. (Chap. 2, Sec. 2, p. 41)

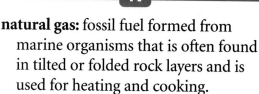

lava: thick, gooey, molten rock material flowing from volcanoes onto Earth's surface. (Chap. 2, Sec. 2, p. 40)

liquefaction: occurs when wet soil acts more like a liquid during an earthquake. (Chap. 5, Sec. 3, p. 141)

lithosphere (LIH thuh sfihr): rigid layer of Earth about 100 km thick, made of the crust and a part of the upper mantle. (Chap. 4, Sec. 3, p. 106)

luster: describes the way a mineral reflects light from its surface; can be either metallic or nonmetallic. (Chap. 1, Sec. 2, p. 16)

magma: hot, melted rock material beneath Earth's surface. (Chap. 1, Sec. 1, p. 11)

magnitude: measure of the energy released during an earthquake. (Chap. 5, Sec. 3, p. 140)

metamorphic rock: forms when heat, pressure, or fluids act on igneous, sedimentary, or other metamorphic rock and affect its form or composition, or both. (Chap. 2, Sec. 3, p. 45)

mineral: naturally occurring inorganic solid that has a definite chemical composition and an orderly internal atomic structure. (Chap. 1, Sec. 1, p. 8)

mineral resources: resources from which metals are obtained. (Chap. 3, Sec. 3, p. 83)

natural gas: fossil fuel formed from marine organisms that is often found in tilted or folded rock layers and is used for heating and cooking. (Chap. 3, Sec. 1, p. 69)

nonfoliated: metamorphic rock, such as quartzite and marble, whose mineral grains grow and rearrange but do not form layers. (Chap. 2, Sec. 3, p. 48)

normal fault: break in rock caused by tension forces, where rock above the fault surface moves down relative to the rock below the fault surface. (Chap. 5, Sec. 1, p. 128)

nuclear energy: alternative energy source that is based on atomic fission. (Chap. 3, Sec. 1, p. 73)

oil: liquid fossil fuel formed from marine organisms that is burned to obtain energy and used in the manufacture of plastics. (Chap. 3, Sec. 1, p. 69)

ore: deposit in which a mineral exists in large enough amounts to be mined at a profit. (Chap. 3, Sec. 3, p. 83)

Pangaea (pan JEE uh): large, ancient landmass that was composed of all the continents joined together. (Chap. 4, Sec. 1, p. 98)

plate: a large section of Earth's oceanic or continental crust and rigid, upper mantle that moves around on the asthenosphere. (Chap. 4, Sec. 3, p. 106)

plate tectonics: theory that Earth's crust and upper mantle are broken into plates that float and move around on a plasticlike layer of the mantle. (Chap. 4, Sec. 3, p. 106)

primary wave: seismic wave that moves rock particles back and forth in the same direction that the wave travels. (Chap. 5, Sec. 2, p. 131)

R

recycling: conservation method in which old materials are processed to make new ones. (Chap. 3, Sec. 3, p. 87)

reserve: amount of a fossil fuel that can be extracted from Earth at a profit using current technology. (Chap. 3, Sec. 1, p. 71)

reverse fault: break in rock caused by compressive forces, where rock above the fault surface moves upward relative to the rock below the fault surface. (Chap. 5, Sec. 1, p. 128)

rock: mixture of one or more minerals, volcanic glass, organic matter, or other materials; can be igneous, metamorphic, or sedimentary. (Chap. 2, Sec. 1, p. 36)

rock cycle: model that describes how rocks slowly change from one form to another through time. (Chap. 2, Sec. 1, p. 37)

S

seafloor spreading: Hess's theory that new seafloor is formed when magma is forced upward toward the surface at a mid-ocean ridge. (Chap. 4, Sec. 2, p. 103)

secondary wave: seismic wave that moves rock particles at right angles to the direction of the wave. (Chap. 5, Sec. 2, p. 131)

sedimentary rock: forms when sediments are compacted and cemented together or when minerals come out of solution or are left behind by evaporation. (Chap. 2, Sec. 4, p. 49)

sediments: loose materials, such as rock fragments, mineral grains, and the remains of once-living plants and animals, that have been moved by wind, water, ice, or gravity. (Chap. 2, Sec. 4, p. 49)

seismic (SIZE mihk) **wave:** wave generated by an earthquake. (Chap. 5, Sec. 2, p. 130)

seismograph: instrument used to register earthquake waves and record the time that each arrived. (Chap. 5, Sec. 2, p. 133)

shield volcano: broad, gently sloping volcano formed by quiet eruptions of basaltic lava. (Chap. 6, Sec. 2, p. 166)

silicate: mineral that contains silicon and oxygen and usually one or more other elements. (Chap. 1, Sec. 1, p. 12)

sill: igneous rock feature formed when magma is squeezed into a horizontal crack between layers of rock and hardens underground. (Chap. 6, Sec. 3, p. 173)

solar energy: energy from the Sun that is clean, inexhaustible, and can be transformed into electricity by solar cells. (Chap. 3, Sec. 2, p. 76)

specific gravity: ratio of a mineral's weight compared with the weight of an equal volume of water. (Chap. 1, Sec. 2, p. 16)

streak: color of a mineral when it is in powdered form. (Chap. 1, Sec. 2, p. 17)

strike-slip fault: break in rock caused by shear forces; where rocks move past each other without much vertical movement. (Chap. 5, Sec. 1, p. 129)

surface wave: seismic wave that moves rock particles up and down in a backward rolling motion and side to side in a swaying motion. (Chap. 5, Sec. 2, p. 131)

T

tephra (TEFF ruh): bits of rock or solidified lava dropped from the air during an explosive volcanic eruption; ranges in size from volcanic ash to volcanic bombs and blocks. (Chap. 6, Sec. 2, p. 166)

tsunami (soo NAHM ee): seismic sea wave that begins over an earthquake focus and can be highly destructive when it crashes on shore. (Chap. 5, Sec. 3, p. 142)

V

vent: opening where magma is forced up and flows out onto Earth's surface as lava, forming a volcano. (Chap. 6, Sec. 1, p. 158)

volcanic neck: solid igneous core of a volcano left behind after the softer cone has been eroded. (Chap. 6, Sec. 3, p. 173)

volcano: opening in Earth's surface that erupts sulfurous gases, ash, and lava; can form at Earth's plate boundaries, where plates move apart or together, and at hot spots. (Chap. 6, Sec. 1, p. 156)

W

wind farm: area where many windmills use wind to generate electricity. (Chap. 3, Sec. 2, p. 77)

Este glosario define cada término clave que aparece en negrillas en el texto. También muestra el capítulo, la sección y el número de página en donde se usa dicho término.

A

asthenosphere / astenosfera: capa viscosa de la Tierra en la cual las placas litosféricas flotan y se mueven. (Cap. 4, Sec. 3, pág. 106)

B

basaltic / basáltica: roca ígnea densa y oscura que se forma del magma rico en magnesio y hierro y deficiente en sílice. (Cap. 2, Sec. 2, pág. 43)

batholith / batolito: cuerpo rocoso ígneo e intrusivo más grande que se forma cuando el magma que es forzado a ascender hacia la corteza terrestre se enfría lentamente y se solidifica bajo tierra. (Cap. 6, Sec. 3, pág. 172)

biomass energy / energía de biomasa: energía renovable proveniente de la quema de materiales orgánicos, como la leña y el alcohol. (Cap. 3, Sec. 2, pág. 79)

C

caldera / caldera: extensa abertura circular que se forma cuando colapsa la parte superior de un volcán. (Cap. 6, Sec. 3, pág. 174)

cementation / cementación: proceso formador de rocas sedimentarias en el cual los cementos naturales que se producen cuando el agua se filtra por la roca y el suelo mantienen unidos los sedimentos grandes. (Cap. 2, Sec. 4, pág. 51)

cinder cone volcano / volcán de cono de carbonilla: volcán de laderas abruptas y vagamente compreso que se forma cuando la tefrita cae al suelo. (Cap. 6, Sec. 2, pág. 166)

cleavage / crucero: propiedad física de algunos minerales de poder romperse a lo largo de superficies suaves y planas. (Cap. 1, Sec. 2, pág. 17)

coal / carbón: roca sedimentaria formada de material vegetal en descomposición; el combustible fósil más abundante en el mundo. (Cap. 3, Sec. 1, pág. 67)

compaction / compactación: proceso que forma rocas sedimentarias cuando las capas de sedimentos pequeños se comprimen debido al peso de las capas superiores. (Cap. 2, Sec. 4, pág. 50)

composite volcano / volcán compuesto: volcán que se ha formado por la alternación de erupciones explosivas y silenciosas que producen capas de tefrita y lava; se halla principalmente donde se juntan las placas terrestres y una placa se hunde debajo de la otra. (Cap. 6, Sec. 2, pág. 167)

continental drift / deriva continental: hipótesis de Wegener que afirmaba que todos los continentes estuvieron unidos en algún momento formando una sola masa continental, la cual se separó hace unos 200 millones de años, haciendo que los continentes derivaran lentamente a sus posiciones actuales. (Cap. 4, Sec. 1, pág. 98)

convection current / corriente de convección: corriente en el manto terrestre que transfiere energía en el interior de la Tierra y que provee la potencia de la tectónica de placas. (Cap. 4, Sec. 3, pág. 111)

crater / cráter: depresión de murallas escarpadas alrededor de la chimenea de un volcán. (Cap. 6, Sec. 1, pág. 158)

crystal / cristal: sólido cuyos átomos están arreglados en un patrón ordenado y repetitivo. (Cap. 1, Sec. 1, pág. 9)

D

dike / dique: filón de roca ígnea que se forma cuando el magma es inyectado en una fisura vertical que atraviesa capas rocosas y se endurece bajo tierra. (Cap. 6, Sec. 3, pág. 173)

E

earthquake / terremoto: vibraciones producidas cuando las rocas se rompen a lo largo de una falla. (Cap. 5, Sec. 1, pág. 127)

epicenter / epicentro: punto sobre la superficie terrestre directamente sobre el foco de un terremoto. (Cap. 5, Sec. 2, pág. 131)

extrusive / extrusiva: roca ígnea de grano fino que se forma cuando el magma se enfría rápidamente sobre o cerca de la superficie terrestre. (Cap. 2, Sec. 2, pág. 41)

F

fault / falla: superficie a lo largo de la cual se mueven y se rompen las rocas cuando exceden su límite de elasticidad. (Cap. 5, Sec. 1, pág. 126)

focus / foco: en un terremoto, es el punto sobre la superficie terrestre donde se libera la energía en forma de ondas sísmicas. (Cap. 5, Sec. 2, pág. 130)

foliated / foliada: roca metamórfica, como la pizarra y el gneiss, cuyos granos minerales se aplanan y se alinean en capas paralelas. (Cap. 2, Sec. 3, pág. 47)

fossil fuel / combustible fósil: recurso energético no renovable, como el petróleo y el carbón, que se formó hace millones de años a partir de los restos de plantas y otros organismos muertos. (Cap. 3, Sec. 1, pág. 66)

fracture / fractura: propiedad física de algunos minerales de romperse a lo largo de superficies disparejas, ásperas o dentadas. (Cap. 1, Sec. 2, pág. 17)

G

gem / gema: mineral precioso, muy valioso que se puede usar en joyería. (Cap. 1, Sec. 3, pág. 19)

geothermal energy / energía geotérmica: recurso energético interminable que hace uso del magma caliente o del calor de las rocas calientes y secas debajo de la superficie terrestre para generar electricidad. (Cap. 3, Sec. 2, pág. 78)

granitic / granítica: roca ígnea de color claro y rica en sílice que es menos densa que la roca basáltica. (Cap. 8, Sec. 2, pág. 43)

H

hardness / dureza: medida del grado de facilidad con que se puede rayar un mineral; se determina según el arreglo de los átomos del mineral. (Cap. 1, Sec. 2, pág. 15)

hot spot / foco caliente: zona de intenso calor ubicada en el límite entre el manto y el núcleo de la Tierra que forma volcanes cuando la roca derretida es forzada a ascender y atraviesa la corteza. (Cap. 6, Sec. 1, pág. 160)

hydroelectric energy / energía hidroeléctrica: electricidad que produce la potencia del agua al hacer grandes represas en un río. (Cap. 3, Sec. 2, pág. 78)

I

igneous rock / roca ígnea: roca intrusiva o extrusiva que se forma cuando el magma caliente se enfría y se endurece. (Cap. 2, Sec. 2, pág. 40)

intrusive / intrusiva: tipo de roca ígnea que, por lo general, contiene cristales de gran tamaño y que se forma cuando el magma se enfría lentamente debajo de la superficie terrestre. (Cap. 2, Sec. 2, pág. 41)

L

lava / lava: materia rocosa derretida, espesa y viscosa, que fluye de los volcanes hacia la superficie terrestre. (Cap. 2, Sec. 2, pág. 40)

liquefaction / liquefacción: ocurre cuando el suelo mojado actúa como un líquido durante un terremoto. (Cap. 5, Sec. 3, pág. 141)

lithosphere / litosfera: capa rígida de la Tierra de unos 100 km de grosor formada por la corteza y parte del manto superior. (Cap. 4, Sec. 3, pág. 106)

luster / lustre: describe la manera en que un mineral refleja la luz desde su superficie; puede ser metálico o no metálico. (Cap. 1, Sec. 2, pág. 16)

M

magma / magma: material rocoso caliente y fundido que se halla debajo de la superficie terrestre. (Cap. 1, Sec. 1, pág. 11)

magnitude / magnitud: medida de la energía liberada durante un movimiento sísmico. (Cap. 5, Sec. 3, pág. 140)

metamorphic rock / roca metamórfica: se forma cuando el calor, la presión o los líquidos actúan sobre rocas ígneas, sedimentarias u otras rocas metamórficas y les afectan la forma o composición o ambas. (Cap. 2, Sec. 3, pág. 45)

mineral / mineral: sólido inorgánico que ocurre en forma natural y que posee una composición química definida y una estructura atómica interna ordenada. (Cap. 1, Sec. 1, pág. 8)

mineral resources / recursos minerales: recursos a partir de los cuales se obtienen metales. (Cap. 3, Sec. 3, pág. 83)

N

natural gas / gas natural: combustible fósil que se formó de organismos marinos y que con frecuencia se encuentra en capas rocosas inclinadas o plegadas y que se usa en calefacción y cocción. (Cap. 3, Sec. 3, pág. 69)

nonfoliated / no foliada: roca metamórfica, como la cuarcita y el mármol, cuyos granos minerales crecen y se reordenan, pero sin formar capas. (Cap. 2, Sec. 3, pág. 48)

normal fault / falla normal: ruptura en la roca causada por las fuerzas de tensión, en donde la roca sobre la superficie de la falla se mueve hacia abajo en relación con la roca debajo de la falla. (Cap. 5, Sec. 1, pág. 128)

Spanish Glossary

nuclear energy / energia nuclear: fuente energética alterna que se basa en la fisión del átomo. (Cap.3, Sec. 1, pág. 73)

oil / petróleo: combustible fósil líquido formado a partir de organismos marinos que se quema para obtener energía y el cual se utiliza en la manufactura de plásticos. (Cap. 3, Sec. 1, pág. 69)

ore / mena: depósito en que un mineral existe en cantidades lo suficientemente grandes para ser minado con fines de lucro. (Cap. 3, Sec. 3, pág. 83)

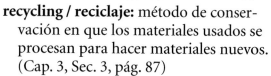

Pangaea / Pangaea: masa de tierra extensa y antigua que una vez estuvo formada por el conjunto de todos los continentes. (Cap. 4, Sec. 1, pág. 98)

plate / placa: región extensa del manto superior rígido y de la corteza oceánica o continental de la Tierra que se mueve sobre la astenosfera. (Cap. 4, Sec. 3, pág. 106)

plate tectonics / tectónica de placas: teoría que afirma que la corteza y el manto superior terrestres se separan en placas que flotan y se mueven sobre una capa viscosa del manto. (Cap. 4, Sec. 3, pág. 106)

primary wave / onda primaria: onda sísmica que mueve las partículas rocosas en un movimiento oscilatorio en la misma dirección en que viaja la onda. (Cap. 5, Sec. 2, pág. 131)

recycling / reciclaje: método de conservación en que los materiales usados se procesan para hacer materiales nuevos. (Cap. 3, Sec. 3, pág. 87)

reserve / reserva: cantidad de combustible fósil que se puede extraer de la Tierra con fines de lucro usando tecnología contemporánea. (Cap. 3, Sec. 1, pág. 71)

reverse fault / falla invertida: ruptura en la roca causada por las fuerzas de compresión, en que la roca sobre la superficie de la falla se mueve hacia arriba en relación con la roca debajo de la falla. (Cap. 5, Sec. 1, pág. 128)

rock / roca: mezcla de uno o más minerales, vidrio volcánico, materia orgánica u otros materiales; puede ser ígnea, metamórfica o sedimentaria. (Cap. 2, Sec. 1, pág. 36)

rock cycle / ciclo de las rocas: modelo que describe el cambio lento de las rocas de una forma a otra a través del tiempo. (Cap. 2, Sec. 1, pág. 37)

seafloor spreading / expansión del suelo marino: teoría de Hess que afirma que el nuevo suelo marino se forma cuando el magma es forzado a subir a la superficie en una dorsal mediooceánica. (Cap. 4, Sec. 2, pág. 103)

secondary wave / onda secundaria: onda sísmica que al moverse hace que las partículas rocosas vibren formando un ángulo recto a la dirección de la onda. (Cap. 5, Sec. 2, pág. 131)

sedimentary rock / roca sedimentaria: se forma cuando los sedimentos se compactan y se cementan o cuando los minerales salen de una solución o cuando la evaporación los deja atrás. (Cap. 2, Sec. 4, pág. 49)

sediments / sedimentos: materiales sueltos, como fragmentos rocosos, granos minerales y restos de plantas y animales, dejados por el viento, el agua, el hielo o la gravedad. (Cap. 2, Sec. 4, pág. 49)

seismic wave / onda sísmica: onda generada por un movimiento sísmico. (Cap. 5, Sec. 2, pág. 130)

seismograph / sismógrafo: instrumento que registra las ondas sísmicas y anota el momento de llegada de cada onda. (Cap. 5, Sec. 2, pág. 133)

shield volcano / volcán de escudo: volcán ancho y de laderas levemente inclinadas que se forma gracias a erupciones silenciosas de lava basáltica. (Cap. 6, Sec. 2, pág. 166)

silicate / silicato: mineral que contiene sílice y oxígeno y, por lo general, uno o más elementos. (Cap. 1, Sec. 1, pág. 12)

sill / intrusión: rasgo rocoso ígneo que se forma cuando el magma es inyectada en una grieta horizontal entre capas de roca y se endurece bajo tierra. (Cap. 6, Sec. 3, pág. 173)

solar energy / energía solar: energía proveniente del sol que es limpia, inagotable y que se puede transformar en electricidad por medio de células solares. (Cap. 3, Sec. 2, pág. 76)

specific gravity / gravedad específica: razón del peso de un mineral comparada con el peso de un volumen igual de agua. (Cap. 1, Sec. 2, pág. 16)

streak / veta: color de un mineral cuando se encuentra en forma de polvo. (Cap. 1, Sec. 2, pág. 17)

strike-slip fault / falla transformante: lugar donde las fuerzas de cizallamiento han ocasionado el rompimiento de las rocas, las cuales se deslizan una al lado de la otra en direcciones opuestas, pero sin mucho movimiento vertical. (Cap. 5, Sec. 1, pág. 129)

surface wave /onda de superficie: onda sísmica que mueve las partículas rocosas de arriba hacia abajo en un movimiento rotatorio y de lado a lado en un movimiento de vaivén. (Cap. 5, Sec. 2, pág. 131)

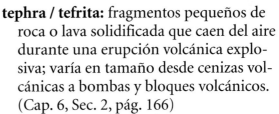

tephra / tefrita: fragmentos pequeños de roca o lava solidificada que caen del aire durante una erupción volcánica explosiva; varía en tamaño desde cenizas volcánicas a bombas y bloques volcánicos. (Cap. 6, Sec. 2, pág. 166)

tsunami / tsunami: onda marina sísmica que comienza sobre el foco de un terremoto y la cual puede ser muy destructiva cuando llega al litoral. (Cap. 5, Sec. 3, pág. 142)

vent / chimenea: abertura por donde sube el magma y sale a la superficie terrestre como lava, formando un volcán. (Cap. 6, Sec. 1, pág. 158)

volcanic neck / chimenea volcánica: núcleo ígneo sólido de un volcán que queda atrás después de la erosión del cono más suave. (Cap. 6, Sec. 3, pág. 173)

Spanish Glossary

volcano / volcán: abertura en la superficie terrestre que arroja gases sulfurosos, cenizas y lava; se puede formar en el límite de las placas terrestres, donde éstas se separan o se juntan y también en los focos calientes. (Cap. 6, Sec. 1, pág. 156)

wind farm / fincas de energía eólica: área en que muchos molinos de viento usan el viento para generar electricidad. (Cap. 3, Sec. 2, pág. 77)

The index for *Earth Materials and Processes* will help you locate major topics in the book quickly and easily. Each entry in the index is followed by the number of the pages on which the entry is discussed. A page number given in boldfaced type indicates the page on which that entry is defined. A page number given in italic type indicates a page on which the entry is used in an illustration or photograph. The abbreviation *act.* indicates a page on which the entry is used in an activity.

Index

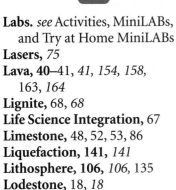

Index

Index

Credits

Art Credits

Glencoe would like to acknowledge the artists and agencies who participated in illustrating this program: Absolute Science Illustration; Andrew Evansen; Argosy; Articulate Graphics; Craig Attebery represented by Frank & Jeff Lavaty; CHK America; Gagliano Graphics; Pedro Julio Gonzalez represented by Melissa Turk & The Artist Network; Robert Hynes represented by Mendola Ltd.; Morgan Cain & Associates; JTH Illustration; Laurie O'Keefe; Matthew Pippin represented by Beranbaum Artist's Representative; Precision Graphics; Publisher's Art; Rolin Graphics, Inc.; Wendy Smith represented by Melissa Turk & The Artist Network; Kevin Torline represented by Berendsen and Associates, Inc.; WILDlife ART; Phil Wilson represented by Cliff Knecht Artist Representative; Zoo Botanica.

Photo Credits

Abbreviation Key: AA=Animals Animals; AH=Aaron Haupt; AMP=Amanita Pictures; BC=Bruce Coleman, Inc.; CB=CORBIS; DM=Doug Martin; DRK=DRK Photo; ES=Earth Scenes; FP=Fundamental Photographs; GH=Grant Heilman Photography; IC=Icon Images; KS=KS Studios; LA=Liaison Agency; MB=Mark Burnett; MM=Matt Meadows; PE=PhotoEdit; PD=PhotoDisc; PQ=PictureQuest; PR=Photo Researchers; SB=Stock Boston; TSA=Tom Stack & Associates; TSM=The Stock Market; VU=Visuals Unlimited.

Cover PD; **vi** Inga Spence/VU; **vii** Larry Ulrich/DRK; **viii** Sigurjon Sindrason; **1** (b)Mark A. Schneider/VU, (t)MB/PR; **2** (t)Jack Smith/AP/Wide World Photo, (b)Gary Stewart/AP/Wide World Photo; **3** (t)Francois Gohier/PR, (b)David Muench/CB; **5** (t)Chlaus Lotscher/SB, (b)USGS; **6** Mark A. Schneider/PR; **6-7** D. Boone/CB; **7** (t)MM, (c)DM, (b)José Manuel Sanchis Calvete/CB; **8** MM; **9** (t)John R. Foster/PR, (b)Mark A. Schneider/VU; **10** (t)Mark A. Schneider/VU, (cl)Albert J. Copley/VU, (cr bl)Harry Taylor/DK Images, (bc)MB, (br)Mark A. Schneider/VU; **11** (l)Patricia K. Armstrong/VU, (r)Dennis Flaherty Photography/PR; **13** KS; **14** (l)MB/PR, (c)Dan Suzio/PR, (r)Breck P. Kent/ES; **15** (t)Bud Roberts/VU, (c)Charles D. Winters/PR, (b)IC; **16** (l)Andrew McClenaghan/Science Photo Library/PR, (r)Charles D. Winters/PR; **17** (t)Geoff Butler, (bl)DM, (br)PR; **18** MM; **19** Reuters NewMedia/CB; **20** (l, t to b)Biophoto Associates/PR, Biophoto Associates/PR, VU, Mark A. Schneider/VU, (r, t to b)H. Stern/PR, A.J. Copley/VU, A.J. Copley/VU, H. Stern/PR; **21** (l, t to b)University of Houston, Arthur R. Hill/VU, DM, DM, (r, t to b)Charles D. Winters/PR, David Lees/CB, A.J. Copley/VU, Vaughan Fleming/Science Photo Library/PR; **22** (l)Francis G. Mayer/CB, (r)Smithsonian Institution; **23** (l)Randolph King/PR, (r)DM; **24** (t)Maurice Nimmo/Frank Lane Picture Agency/CB, (bl)Paul Silverman/FP, (br)Biophoto Associates/PR; **25** Jim Cummins/FPG; **26** (t)DM, (c)Jose Manuel Sanchis/CB, (b)MM; **27** (l)Charles D. Winter/PR, (r)Andrew J. Martinez/PR; **28** SPL/Custom Medical Stock Photo; **30** (tl)Mark A. Schneider/PR, (tr)Charles R. Belinky/PR, (c)Phillip Hayson/PR, (b)Mark A. Schneider/PR; **32** Arthur Hill/VU; **34** John D. Cunningham/VU; **34-35** Cliff Leight; **35** Geoff Butler; **36** (l)CB, (r)DM; **37** (tl)Steve Hoffman, (tr)Breck P. Kent/ES, (bl)Brent Turner/BLT Productions, (br)Breck P. Kent/ES; **38** (bkgd)CB/PQ, (t)CB, (bl)Martin Miller, (bc)Jeff Gnass, (br)Doug Sokell/TSA; **39** Russ Clark; **40** USGS/HVO; **41** (t)Breck P. Kent/ES, (b)DM; **42** (cl)Mark Steinmetz, (cr bl)DM, (br)Tim Courlas, (others)Breck P. Kent/ES; **44** (l)Breck P. Kent/ES, (r)DM/PR; **45** (tl)Breck P. Kent/ES, (tc)courtesy Kent Ratajeski & Dr. Allen Glazner, University of NC, (tr)Breck P. Kent/ES; **47** (l)John Evans, (r)Robert Estall/CB; **48** Paul Rocheleau/Index Stock; **49** (l)Timothy Fuller, (r)Steve McCutcheon/VU; **51** (l)IC, (cl)DM, (cr)Andrew Martinez/PR,

(r)John R. Foster/PR; **52** (l)Breck P. Kent/ES, (r)AH; **53** (t)Georg Gerster/PR, (b)IC; **55** Beth Davidow/VU; **56** (t)Breck P. Kent/ES, (c)IC, (bl)Jack Sekowski, (br)Tim Courlas; **58-59** Y. Kawasaki/Photonica; **59** Matt Turner/LA; **60** (t)Mark Segal/Index Stock, (c)Albert J. Copley/VU, (b)DM; **62** Jeremy Woodhouse/DRK; **64** Lowell Georgia/CB; **64-65** Bill Ross/CB; **65** MM; **67** VU; **70** (l)George Lepp/CB, (r)Carson Baldwin Jr./ES; **71** Paul A. Souders/CB; **72** (tl)Emory Kristof, (tr)National Energy Technology Laboratory, (b)Ian R. MacDonald/Texas A&M University; **73** Hal Beral/VU; **75** Roger Ressmeyer/CB; **76** Spencer Grant/PE; **77** Inga Spence/VU; **78** Robert Cameron/Stone; **79** Vince Streano/CB; **80** (t)David Young-Wolff/PE, (b)Earl Young/Archive Photos; **81** Peter Holden/VU; **83** AH; **84** Joseph Nettis/PR; **85** (t)Mark Joseph/Stone, (bl)AH, (br)Wyoming Mining Association; **88** (t)Joel W. Rogers/CB, (b)AH; **89** AH; **90** Brown Brothers; **91** (l)Ed Clark, (r)Shell Oil Co.; **92** (t)Betty Sederquist/VU, (c)Bob Rowan/CB, (b)C. Osborne/PR; **93** (tl)Michael Mancuso/Omni-Photo Communications, (tr)file photo, (bl)Andrew J. Martinez/PR, (br)Coco McCoy/Rainbow; **94** David Frazier; **96** Francois Gohier/PR; **96-97** Stone; **97** MB; **100** Martin Land/Science Source/PR; **103** Ralph White/CB; **109** Davis Meltzer; **110** Craig Aurness/CB; **112** Craig Brown/Index Stock; **113** Ric Ergenbright/CB; **114** Roger Ressmeyer/CB; **116** Burhan Ozbilici/AP/Wide World Photos; **118** L. Lauber/ES; **119** Courtesy Ed Klimasauskas; **120** Galen Rowell/CB; **121** (l)Tim Barnwen/SB, (r)Bettmann/CB; **124** Bettmann/CB; **124-125** Paras Shah/AP/Wide World Photo; **125** MB; **126** Tom & Therisa Stack/TSA; **128** (t)Tom Bean/DRK, (b)Lysbeth Corsi/VU; **129** David Parker/PR; **130** Tom & Therisa Stack/TSA; **132** Robert W. Tope/Natural Science Illustrations; **139** (l)Steven D. Starr/SB, (r)Berkeley Seismological Laboratory; **140** Hurriyet/AP/Wide World Photo; **141** David J. Cross/Peter Arnold, Inc.; **144** James L. Stanfield/National Geographic Society; **145** Courtesy Safe-T-Proof; **146** (l)Ben Simmons/TSM, (r)Reuters NewMedia/CB; **148** (l)Bettmann/CB, (r)RO-MA Stock/Index Stock; **149** Richard Cummins/CB; **150** (l)Reuters/STR/Archive Photos, (r)Russell D. Curtis/PR; **151** (l)Science VU/VU, (r)Peter Menzel/SB; **152** Vince Streano/CB; **154** Roger Ressmeyer/CB; **154-155** Fabrizio Villa/AP/Wide World Photos; **155** KS; **156** Sigurjon Sindrason; **157** (t)John Cancalosi/DRK, (b)Deborah Brosnan, Sustainable Ecosystems Institute; **161** NASA/CB; **162 163** Gary Rosenquist; **164** (bkgd)API/Explorer/PR, (t)G. Brad Lewis/Photo Resource Hawaii, (bl)Robert Hessler/Planet Earth Pictures, (br)Paul Chesley; **165** (l)Steve Kaufman/DRK, (r)Dee Breger/PR; **166** James L. Amos/CB; **167** (t)Krafft/Explorer/Science Source/PR, (b)Darrell Gulin/DRK; **169** Bernard Edmaier/PR; **172** (t)Joyce Photo/PR, (bl)Brent Turner, (br)DM; **173** (tl)Dick Canby, (tr)Tom Bean/DRK, (b)Jeff Foott/TSA; **175** Larry Ulrich/DRK; **176** AMP; **177** Spencer Grant/PE; **178** (bkgd)Mimmo Jodice/CB, (inset)Roger Ressmeyer/CB; **179** Jonathan Blair/CB; **180** (tl)Dave B. Fleetham/TSA, (tr b)Tom Bean/DRK; **181** (l)Soames Summerhays/PR, (r)Photri/TSM; **186-187** PD; **188** Garry D. McMichael/PR; **188-189** Harvey Wood/The Still Moving Picture Co.; **189** (t c)AH, (b)Daniel Chester French/CB; **190** (t)SuperStock, (bl)Lindsay Hebberd/CB, (br)PhotoTake NYC/PQ; **191** (t)Ric Ergenbright/CB, (bl)IC, (br)Raymond Gehman/CB; **192** Michell D. Bridwell/PE; **196** David Young-Wolff/PE; **198** Kaz Chiba/PD; **199** Dominic Oldershaw; **200** StudiOhio; **201** MM; **203** (tl tr c)Mark Steinmetz, (bl)Elaine Shay, (br)Brent Turner/BLT Productions; **206** Paul Barton/TSM; **209** AH; **209** Bettmann/CB.

Acknowledgments

"Listening In" by Gordon Judge. Reprinted by permission of the author.

Credits

PERIODIC TABLE OF THE ELEMENTS

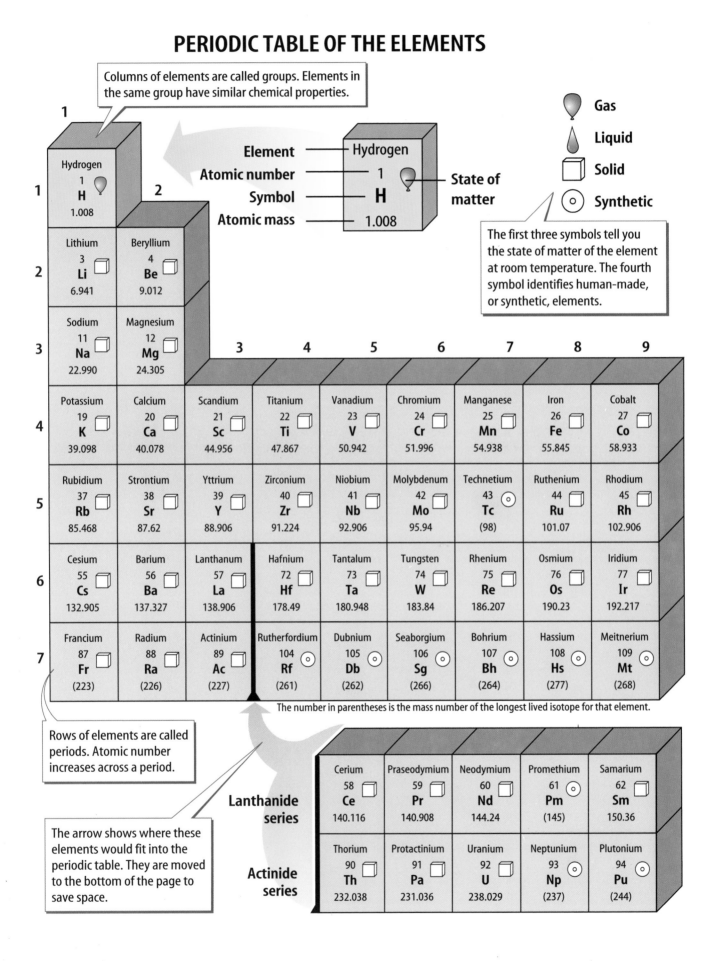

Columns of elements are called groups. Elements in the same group have similar chemical properties.

Gas
Liquid
Solid
Synthetic

Element — Hydrogen
Atomic number — 1
Symbol — H
Atomic mass — 1.008
State of matter

The first three symbols tell you the state of matter of the element at room temperature. The fourth symbol identifies human-made, or synthetic, elements.

1	2	3	4	5	6	7	8	9
Hydrogen 1 **H** 1.008								
Lithium 3 **Li** 6.941	Beryllium 4 **Be** 9.012							
Sodium 11 **Na** 22.990	Magnesium 12 **Mg** 24.305							
Potassium 19 **K** 39.098	Calcium 20 **Ca** 40.078	Scandium 21 **Sc** 44.956	Titanium 22 **Ti** 47.867	Vanadium 23 **V** 50.942	Chromium 24 **Cr** 51.996	Manganese 25 **Mn** 54.938	Iron 26 **Fe** 55.845	Cobalt 27 **Co** 58.933
Rubidium 37 **Rb** 85.468	Strontium 38 **Sr** 87.62	Yttrium 39 **Y** 88.906	Zirconium 40 **Zr** 91.224	Niobium 41 **Nb** 92.906	Molybdenum 42 **Mo** 95.94	Technetium 43 **Tc** (98)	Ruthenium 44 **Ru** 101.07	Rhodium 45 **Rh** 102.906
Cesium 55 **Cs** 132.905	Barium 56 **Ba** 137.327	Lanthanum 57 **La** 138.906	Hafnium 72 **Hf** 178.49	Tantalum 73 **Ta** 180.948	Tungsten 74 **W** 183.84	Rhenium 75 **Re** 186.207	Osmium 76 **Os** 190.23	Iridium 77 **Ir** 192.217
Francium 87 **Fr** (223)	Radium 88 **Ra** (226)	Actinium 89 **Ac** (227)	Rutherfordium 104 **Rf** (261)	Dubnium 105 **Db** (262)	Seaborgium 106 **Sg** (266)	Bohrium 107 **Bh** (264)	Hassium 108 **Hs** (277)	Meitnerium 109 **Mt** (268)

The number in parentheses is the mass number of the longest lived isotope for that element.

Rows of elements are called periods. Atomic number increases across a period.

The arrow shows where these elements would fit into the periodic table. They are moved to the bottom of the page to save space.

Lanthanide series

Cerium 58 **Ce** 140.116	Praseodymium 59 **Pr** 140.908	Neodymium 60 **Nd** 144.24	Promethium 61 **Pm** (145)	Samarium 62 **Sm** 150.36

Actinide series

Thorium 90 **Th** 232.038	Protactinium 91 **Pa** 231.036	Uranium 92 **U** 238.029	Neptunium 93 **Np** (237)	Plutonium 94 **Pu** (244)